Debates in Economic History

Edited by Peter Mathias

Technological Change: The United States and Britain in the Nineteenth Century

Technological Change: The United States and Britain in the Nineteenth Century

edited with an introduction by
S. B. SAUL

METHUEN & CO LTD
11 NEW FETTER LANE LONDON EC4

First published in 1970 by Methuen & Co Ltd
Introduction © 1970 by S. B. Saul
Printed in Great Britain by
Richard Clay (The Chaucer Press), Ltd,
Bungay, Suffolk

SBN 416 14730 5 hardback
SBN 416 27890 5 paperback

Distributed in the U.S.A.
by Barnes & Noble Inc.

Contents

Preface

An up-to-date technology, with a continuing capacity to promote and accept technical change, is the essential hall-mark of a modern economy and the most important single agency in effecting higher productivity. Hence, effectively diffusing modern technology to developing countries, and promoting their capacity for absorbing it, is crucial if they, like the present advanced economies in earlier generations, are to break out of a prison of poverty and standards of manual labour penal to the point of servitude. For 'leader economies' processes of invention, innovation, and the diffusion of techniques (the lags associated with the latter process being reflected in the gap between best-practice technology and average technology in an industry) can only be understood in terms of the fundamental characteristics of the economies concerned. Equally, for 'follower economies', sustaining technical change is not simply a matter of importing new machines. If imported technology is to be successfully transplanted, many aspects of the new context have to be favourable – ranging from market opportunities for output to essential 'back-up' skills in a wide arc of supporting functions. The interdependence is wide. Tractors rusting in peasant fields for want of maintenance skills or local repair shops and new power stations offering energy to absent industrial plants symbolize the interrelatedness. 'Behind productivity', remarked W. E. G. Salter, 'lie all the dynamic forces of economic life: technical progress, accumulation, enterprise, and the institutional pattern of society.'

This book takes as its theme a central case-study of the diffusion of technical change among advanced economies in the nineteenth century. Recent work in economic theory, notably enhanced by Dr Salter's book *Productivity and Technical Change* (1960), has been developing greater precision – and greater complexity – to conceptual formulations in this field. Stimulated by Mr Habakkuk's work, as all economic historians working on these problems must be, much research is now in progress to apply these new standards of analysis to the more intractable media of historical evidence. Professor Saul's introduction demonstrates the complexities of the counter-flows and cross-currents which make up the stream of technical change. As so often in contemporary historical analysis, at least in economic history, unitary explanations and simple assumptions in terms of brilliant inventors (or, in these latter days, high investment in applied science) or different relative costs and availability of labour, capital, and resources are not sufficient. Even the conceptually clear difference between adopting different combinations of factors within a known technology and enjoying different propensities to invent new technologies becomes disarmingly difficult to interpret in practice. But the net result is not in doubt and it must give cause for thought to all observers in both advanced and developing countries alike. In the eighteenth century Britain became the fount of the advanced technology which gave industrialization to the world, including the New World; by the twentieth century the main flow of new technology, if not of invention, was from the new world to the old and a very substantial gap in productivity had opened between the United States and the most advanced European economies. In the exploration of this gap lies a world of relevance for explaining the springs of technical progress.

All Souls College, Oxford   PETER MATHIAS

12 April 1970

Acknowledgements

The editor and publishers wish to thank the following for permission to reproduce the articles listed below:

Howard L. Blackmore and the Gun Digest Publishing Company for 'Address Col. Colt, London' (retitled here: 'Colt's London Armoury'), by Howard L. Blackmore (*Gun Digest*, 1958); D. L. Burn for 'The Genesis of American Engineering Competition, 1850–1870' (*Economic History*, Vol. II, January 1931); Cambridge University Press for extracts from *American and British Technology in the Nineteenth Century*, by H. J. Habakkuk (C.U.P., 1962), reproduced here under the title 'The Economic Effects of Labour Scarcity'; The Royal Economic Society for 'The Enfield Arsenal in Theory and History', by N. Rosenberg and E. Ames (*Economic Journal*, December 1968); Professor L. Sandberg and *The Quarterly Journal of Economics* for 'American Rings and English Mules' (*Quarterly Journal of Economics*, February 1969); and Professor S. B. Saul for 'The Market and Development of the Mechanical Engineering Industries in Britain, 1860–1870' (*Economic History Review*, Vol. XX, 1967).

Editor's Introduction

Since the early decades of the nineteenth century a succession of informed observers has commented upon the differences between British and American industrial methods. By the end of the century labour productivity in American industry as a whole was about twice that in Britain and by the 1960s it was more than three times as great. According to Denison, in 1960 national income per person employed in non-agricultural industry in Britain was in real terms only 57 per cent of that in the United States.[1] Duncan Burn's article, reprinted here, was one of the first to examine and document the special features of American technology that were emerging before the Civil War and it remains a major contribution to the subject. Since then there have been several other important studies and this selection of readings concentrates on one of these – the controversial and stimulating interpretation of H. J. Habakkuk in his book *American and British Technology in the Nineteenth Century*. His method was to trace the origins of these differences, to elucidate the theoretical issues involved, and to look at them generally in the light of current historical knowledge. The book deliberately concentrated on expounding ideas rather than upon presenting the fruits of detailed research. The extract reprinted is taken from the first three chapters which analyse American rather than British technology. The other readings, with the exception of the short piece by Blackmore, show some of the fruits of Habakkuk's work, as different scholars have probed further into the ideas and issues that he posed. The aim of this introduction is to set out the basic elements of Habakkuk's arguments and then to point out some of the main stumbling blocks in his interpretation.

[1] Edward F. Denison, *Why Growth Rates Differ* (Washington, D.C., 1967), p. 219.

II

Habakkuk poses a dual problem: what determined the nature and the speed of technological change in the United States and why was Britain's response so tardy in several critical sectors? He argues that American industrial development after 1870 was due, in the main, to her abundant natural resources and the very fast growth of market demand. But the basic elements of the technology which was exploited so rapidly after the Civil War were established earlier. This is his main concern from the American point of view, and in a subsequent comment he has narrowed the crucial period down to the years preceding the arrival of the first great surge of immigrants in the late 1840s.[1]

This technology, he argues, reflected two fundamental economic relationships:

(*a*) The cost of labour relative to that of capital was higher in the United States: more specifically, the ratio of the wage rate, to the price of machines multiplied by the rate of interest plus the rate of depreciation, was higher.

(*b*) The supply of labour in the United States was less elastic. A manufacturer's desire to extend his activities, to widen his capital stock, to employ the same techniques more extensively, was more likely to come up against steeply rising labour costs in the United States and this led to a desire to deepen capital, to use more capital-intensive methods to offset the fall in marginal profit rates.

Elsewhere Habakkuk put it another way.[2] The relative abundance of labour in Britain meant less restraint upon profit and so upon capital accumulation; the opposite was the case in the United States. The need to avoid a falling rate of profit, he argued, is more compelling than the urge to raise that rate. Also, a capital shortage is less likely to be a continuous stimulus to innovation than a labour shortage, the supply of finance being more directly responsive to a successful innovation than the supply of labour. Dear labour compelled

---

[1] H. J. Habakkuk, 'Second Thoughts on British and American Technology in the Nineteenth Century', *Business Archives and History*, III (1963), p. 192.

[2] Ibid., pp. 187–94.

the American manufacturers to make a more careful and systematic investigation of capital-using possibilities and so they adopted earlier, mechanical methods which would have been profitable in Britain.

Habakkuk therefore sees the differences between the two countries largely in terms of factor prices and the faster growth of demand in the United States. He does not ignore other influences such as the inertia of British employers and the hostility of labour in Britain, but he prefers to concentrate on rational responses to economic variables.

The process was intensified by two factors. First, the technical possibilities were greater at the capital-intensive level. For example, the desire for new types of machines led to the creation of an active machine-tool industry. The ability to make one machine made it easier to develop others and the techniques of making machines and final products interchanged. It is the process now known as 'technological spin-off'. Secondly, for a variety of reasons, the rate of creation of new capacity was higher in the United States than in Britain. Apart from the greater direct opportunities this offered for the incorporation of new techniques, productivity grew faster and this meant that the price of commodities, including machines, fell in relation to the cost of labour faster than in Britain, thereby furthering the tendency for the Americans to use capital intensive techniques.

If all this was so, how did the prevailing factor relationship become established? Habakkuk's answer is the traditional one that the high supply price of labour was set by the attractive alternatives offered in agriculture. Evidence from the north-eastern states that by 1850 demand for labour for factories and construction projects was pushing up the price of farm labour, is explained away as reflecting an early shift away from agriculture in those regions as the frontier moved westwards.[1] This may be, but there is other evidence that in boom periods construction was having the same effect on the cost of labour – particularly unskilled labour – in the West too.[2] The problem

---

[1] H. J. Habakkuk, *American and British Technology in the Nineteenth Century* (Cambridge, 1962), pp. 41–2.

[2] P. A. David, 'The Mechanization of Reaping in the Ante-Bellum Mid-West', in H. Rosovsky, *Industrialization in Two Systems* (New York, 1966), p. 15, note 23.

is a crucial one, for it is hardly possible to see the growth of American technology through a simple relationship of factor prices if these were already being determined by the nature and pace of industrial growth.

Such is the core of the argument. The basic difficulty lies in its very general nature. It is never quite clear precisely what it is that has to be explained. Certainly, contemporaries made frequent reference to the dearness of American labour but in itself that does not take us very far. How much more expensive was it and how did this vary between different types of labour? Was it in fact all that expensive anyway, if account is taken of the higher education level, the longer hours worked, the faster pace of work, the intelligent interest in, and contribution to, solving technical difficulties that was said to be so typical of the American worker and that Habakkuk goes out of his way to stress? What was the relative price of capital in Britain compared with different regions in the United States and how far would this indicate the availability of capital anyway? Did the British always use less capital-intensive techniques and neglect American innovations over the whole industrial spectrum? Is it innovation or the diffusion of innovations that concerns us more?

Apart from these general questions, there are others of a more specific kind. It is elementary analysis that a distinction must be drawn between a shift of factor proportions within a given technology and a change in that technology itself. The former implies that in the current state of knowledge, a range of combinations of capital and labour can be used to produce a given output and the combination chosen depends on the relative prices of capital and labour. The latter implies the establishment of a new range of combinations. The point is absolutely fundamental but without the kind of information that rarely comes the way of historians, it is difficult to make the distinction in practice. As Salter said, both types of change take place simultaneously and consequently it is extremely difficult to distinguish how far new techniques are the product of new knowledge or of shifting factor prices.[1] A decision not to use an American combination of factors may be perfectly rational where different factor prices exist, unless the new tech-

[1] W. E. G. Salter, *Productivity and Technological Change* (Cambridge, 1960), p. 20.

nique saves both labour and capital, in which case it would obviously be worth while in Britain too. Decisions to adopt or not to adopt a new technology require quite different explanations. Habakkuk is aware of these distinctions; most of his argument seems to be couched in terms of shifting factor proportions, though he asserts that the search for new technologies had a labour-saving bias too. This is a crucial issue and we have already mentioned one argument Habakkuk puts forward in its support. All the same, apart from purely technological considerations of that kind, there is no theoretical reason why, in developing new techniques, an entrepreneur should wish to save labour costs more than capital costs: his interest is in reducing total costs.

An interesting alternative approach to these problems has been put forward by Stanley Lebergott. He argues that no entrepreneur ever knows for certain how factor prices will move in the future. In nineteenth-century America social mobility – not necessarily the pull of agriculture – created an incessant pressure towards higher wages, so that, with the future price of capital uncertain, the wisest choice was to adopt techniques which were not labour intensive.[1] He also suggests that many entrepreneurs had just left Europe and to them wages seemed high in an absolute sense and they acted upon that feeling. It is not clear that this is a very persuasive argument for the years prior to 1850, but Habakkuk makes a similar point when he argues that, with rudimentary accounting techniques, employers knew their labour costs accurately but had only the haziest ideas of their capital costs.[2] They therefore had a greater tendency to save on the known factor in the United States than in Britain.

The analysis by factor proportions faces the well-known difficulty that since capital goods are made with labour, higher labour costs result in the same proportionate increase in the price of capital goods and so provide no inducement to substitute capital for labour. The problem can be overcome if the American manufacturer, facing high labour costs, buys

[1] S. Lebergott, *Manpower in Economic Growth* (New York, 1964), pp. 230–1. See also M. M. Postan, *Economic History of Western Europe* (Cambridge, 1961), p. 168 for a similar emphasis in a different context.

[2] H. J. Habakkuk. See p. 39 of this volume – pp. 26–7 in the original book.

machines made by lower-paid British labour, but this is hardly a relevant explanation if we are trying to understand the emergence of a distinctive American technology. Habakkuk argues that the skilled labour used to make machines in America was in less scarce supply than the unskilled labour used to operate them. Subsequent research confirms the high premium on the less-skilled workers in the United States – partly because the skilled were so highly paid in Britain – but it is open to question whether in practice any but the most skilled could have operated successfully the crude machine tools used before 1850.[1]

An alternative solution is to introduce a third factor into the analysis. Rosenberg has shown that most of the early unique American innovations were concerned with the cutting and shaping of wood – Blanchard's stocking lathe is but the best known of them.[2] Though they were labour-saving, they were extremely wasteful of wood. However, as wood was much cheaper in the United States than in Britain, the new wood-using, labour-saving devices were preferred there.

III

The making of rifles and revolvers by interchangeable techniques was the most famous of the early American technologies. Some of the analytical problems involved are taken up in an article by Ames and Rosenberg, while another by Howard Blackmore examines Colt's short-lived attempt to make revolvers by these methods in London. Both are reprinted in this volume. How did the technology emerge? Why did Britain lag? What part did labour costs play? Did the British know of the new methods and reject them? What were the roles of government and of the private market? Taking the last point first, Ames and Rosenberg stress that an interchangeable weapon was an entirely different commodity from a hand-made one. A British purchaser of a sporting rifle wanted high-quality finish and a gunstock made to fit. Unlike the United States, there has never been any private demand in Britain for mass-produced interchangeable revolvers or rifles. For military purposes such weapons had obvious advantages, but during

[1] See generally N. Rosenberg, 'Anglo-American Wage Differences in the 1820s', *Journal of Economic History*, XXXVII (1967).

[2] N. Rosenberg (ed.), *The American System of Manufactures* (Chicago, Ill., 1969), p. 76.

the years of peace between Waterloo and the Crimean War the British Government gave no guarantees of regular orders which would have encouraged the private makers in Birmingham to change their techniques and organization. The American Government, on the other hand, subsidized Whitney and others in their experiments. Colt's experience in London shows how dangerous the market problems could be without some kind of official support.

A closer look at the early American experiments in firearms makes one wonder if labour-saving was a very real element at that stage, though it may have been critical for the later extension and diffusion of these methods. The idea was not new: it had already been applied to the making of muskets in France. The main advantage sought was not a cheaper weapon but one that could be more easily serviced. We know very little about the machinery used in the early American armouries but, as Woodbury points out, the rapid rise in the number of skilled armourers employed raises some doubts about the inelasticity of the labour supply.[1] In the early developments, actual interchangeability was achieved, when it was, through laborious filing down of metal to patterns and gauges. It was obtained in some branches of heavy engineering and in the making of cotton spindles in Britain prior to 1850 in this same way. The ultimate American triumph was to produce special machines to replace these hand methods and here, as many have pointed out, lies the key to the American system, not the products themselves. But we know all too little about the forces motivating these superb technical feats. The critical idea of using a rotating cutter for milling was demonstrated in Britain by Bodmer as early as 1824, but with odd exceptions the miller proved of little use in the typical heavy engineering shop in Britain until the cutters were greatly improved several decades later. In the United States itself progress was also slow initially. Fitch tells us that at one of the big private armouries in 1839 there was only one milling machine and at Springfield, the very centre of innovation, they were still using six men to each machine. 'It was some fifteen years later before the manufacture of milling, edging and other important gun machinery was

[1] R. S. Woodbury, 'The Legend of Eli Whitney and Interchangeable Parts', *Technology and Culture*, I (1960), pp. 242–3.

conducted on a scale sufficiently extensive for the general outfitting of the large armories.'[1] As it was there were constant problems in the manufacturing processes, so much so that at Springfield in 1844–9 interchangeable manufacture was to all intents and purposes abandoned and the same was true of the Colt factory.[2] Ignoring the market problems already mentioned, and setting aside the superb stocking machine, we might reasonably suggest that the British arms makers who were not adopting American techniques prior to 1850, were not missing much. The golden rule of mass production is not to incur the huge overhead expenses of tooling up until you are reasonably confident that the product and the technique is right. This is not to say that the British closely monitored all that was going on, yet, when a demand for new weapons arose, they knew where to look and some British machine-tool makers, at least, were prepared to supply some of the machines.

Except for the wood-working machines, it is hard to see either the American experiments or the British response as fitting any rigid factor price mould. In that highly skill-intensive area of New England, where the call of agriculture was less pressing, with a government anxiously subsidizing the experiments, is it not reasonable to believe that the main motivating force in creating the new machine tools was to achieve *technically* the interchangeability that was so clearly escaping them? It is one thing to argue that had the machines been prodigal of labour they would not have been widely used or have called for further modification: it is quite another to suggest that the high labour costs initiated the search for, or basically determined the nature of, those machines.[3] The impact of

[1] C. H. Fitch, *Report on the Manufacture of Interchangeable Parts* (U.S. Census, 1880), p. 5.

[2] F. J. Deyrup, *Arms Makers of the Connecticut Valley* (Northampton, Mass., 1948), p. 145.

[3] The complexities of this analysis of gun making may be somewhat reduced by taking account of Nordhaus's argument that the bias towards labour-saving technological change showed itself more in improvements of products and processes than in basic inventions. In the early days of a product's development, it was often easier to use the more flexible input of labour and only when the basic process had been understood and the product's future assured, could machines be substituted. W. D. Nordhaus, *Invention, Growth and Welfare* (Massachusetts Institute of Technology, 1969), p. 13.

labour costs becomes even less sure when one recalls the two-way nature of the flows of technology. In the decade that saw American machinery being imported into Enfield, British machinery and skilled operators were being brought to the United States to introduce their barrel-rolling process. For the rifling of barrels the British hook-rifling method was adopted by Remingtons in 1861 and soon became the most widely used technique.[1] It is perhaps significant that neither of these processes was in any way aided by the technical possibilities offered by the new American machining methods.

IV

As a contrast we can look at an example of interchangeable manufacture in Britain, almost contemporaneous with the American gun-making experiments, and sometimes described as matched only by them. This was the construction of the Crystal Palace for the Exhibition of 1851. The uniqueness of the building has long been recognized by architects and is described by Rosenberg though he fails to draw out the crucial factors determining its construction.[2] The building was divided into a system of small prefabricated units – the wooden ridge and furrow frames for the roof, the iron lattice girders on which these and the panes rested and the cast-iron supporting pillars. An ingenious machine invented by Paxton was used for grooving, cutting, and finishing the wooden sash-bars – 205 miles of them; the one machine, operated by a man and a boy, did 9 miles a week.[3] The metal parts were manufactured away from the site and erected without further on-the-spot machining or filing. It was a magnificent break-through in design, in mass production, and in assembly techniques. Why did it emerge at this time? Paxton had already tried out some of the techniques on a much smaller building, the Great Conservatory at Chatsworth in 1838, but above all it was the requirement that determined the mode of construction. The building, going up in Hyde Park, had to be taken down and re-erected elsewhere and it was needed in a great hurry. The

[1] C. H. Fitch, op. cit., p. 11.

[2] See Henry Russell Hitchcock, *Early Victorian Architecture in Britain* (London 1954), Vol. I, pp. 540–1, and also N. Rosenberg, *American System of Manufactures*, p. 5.

[3] *Illustrated London News*, 23 November 1850, p. 407.

agreement to build was not signed until 16 July 1850 and, almost incredibly, the building was handed over complete on 31 January 1851. Both of these demands could only be met by a prefabricated interchangeable construction and a tightly controlled system of assembly. Why, then, was the method not employed elsewhere? Fox and Henderson, the contractors, also built the Midland Railway Station at Oxford by the same techniques. Their tender was so far below that of other contractors that it is surprising that other stations were not built to this standard pattern.[1] Possibly the Victorian preoccupation with style took attention away from new methods of construction. There was something of a vogue for prefabricated iron buildings in the 1850s and a considerable export trade in houses and churches constructed in the same way. The technique failed to develop, not for technological or cost reasons, but rather because the product was not in demand except under very particular circumstances.

Interchangeability was not uniquely American; its adoption was not necessarily linked with high labour costs. It often meant the manufacture of a different, standardized, sometimes lower quality product. The major American achievement came through the machine tools that made technically possible the light machining of parts to the required accuracy. But a distinction has to be drawn between the development of the means of manufacturing guns, sewing machines, reapers, and the like, and the circumstances in which these objects, some of which were distinctly labour-saving in themselves, emerged.

Taking the sewing machine as an example, undoubtedly it saved labour in the garment and shoe-making trades. Yet so obvious were these potential savings that many efforts were made to produce such a machine in Europe for a century before Howe was able to perfect one from the basic technical principles established by the earlier inventors. Thimmonier built and operated successfully several sewing machines in France, making army clothing, before they were destroyed by a mob in 1841. The early success of the sewing machine in the United States was partly due to the existence in the east of a ready-made clothing industry, supplying slops to sailors on leave and cheap clothes for slaves. Cheap unskilled labour was required

[1] H. R. Hitchcock, op. cit., p. 540.

to operate the machines and by chance their invention coincided with the arrival in Massachusetts of the cheap Irish labour that made the creation of factory garment and shoe-making industries possible. Each depended on the other for the successful establishment of the industries but neither was called into being by the other's existence. It was the absence of this labour surplus which hindered the emergence of such industries in New York where the making of ready-made clothing had been even more advanced prior to 1845. All this was in sharp contrast to the position in Britain where demand for ready-made clothes appears to have been slight.

The exploitation of the sewing machine in the United States was a function of the market structure and of a ready supply of labour for the using industries. Manufacture of the sewing machines by the new machine tools was the final step, though it was almost entirely a post-Civil War phenomenon in view of the low level of output and frequent model changes experienced before that time. But it is significant that once Singers had established a factory in Scotland, using these same manufacturing techniques and making the machine tools as well as the sewing machines there, it soon developed into the largest factory of its kind in the world. Singer stockholders were told in 1885 that unit costs were thirty per cent below those in the New York factory.[1]

These complex patterns affecting the emergence of the sewing machine industry, raise the question of why Britain did not make more use of the new American ideas after 1860. Why did the abundance of labour in Britain blunt the incentive for British entrepreneurs to use labour-saving devices in a way it did not in the United States nor, as the Singer experience showed, for American entrepreneurs operating under British factor supply conditions? With the heavy migration of unskilled labour away from agriculture, the labour supply position was not unlike that in the United States. Indeed, it is the failure of this abundant labour supply to have much apparent influence on productivity which is one of the most puzzling features of British growth. Habakkuk's explanation is couched mainly in terms of institutional or aggregate

[1] R. B. Davies, 'The Singer Manufacturing Company in Foreign Markets', *Business History Review*, XLIII (1969), pp. 316–17.

demand factors. It could hardly be otherwise, given the success of some enterprises that directly and quickly imitated American manufacturing patterns.

Habakkuk's analysis of the American response to cheap labour after 1845 in the East, and post-1865 elsewhere, is that it was skilled rather than general labour that was now in shortest supply and the incentive was less to save labour, than to switch it from skilled to unskilled, though the highly mechanized methods would probably have been the most profitable even had unskilled labour been less abundant. 'The principal effect of immigration was not on the manufacturer's choice of techniques but on his ability to give effect to his choice'.[1] Though this is almost certainly true for many industries, McGouldrick has suggested another explanation for the impact of immigration upon cotton spinning.[2] He argues that the coming of cheap labour led to the use of more capital per worker because the utilization of less skilled workers forced formerly complex jobs to be broken down into stages requiring different skill levels. But such a decomposition of jobs required more machines per worker because the substitution of unskilled for skilled labour necessitated the slower running of machinery. The effect is much the same: use of the new elastic supply of labour brought either a shift in factor proportions in the form of slowing the pace of existing machines, or a new technique that immigrants could use, such as the sewing machine.

An oddity in Habakkuk's analysis is the typewriter, which he mistakenly places among the pre-1860 innovations in the United States. In fact it was not patented until 1868 and manufactured commercially only from 1873. The point here is that although, as arms makers, Remingtons quickly adapted the interchangeable techniques to the making of typewriters, very few were sold and they gave up the business in 1886.[3] The 1890 census showed only 1,735 workers making them throughout the country. Possibly one reason for the delay was the necessity to establish schools to train clerks to use the machines effectively. In any case the long gap between inven-

[1] H. J. Habakkuk, *American and British Technology*, p. 131.

[2] P. E. McGouldrick, *New England Textiles in the Nineteenth Century* (London, 1968), p. 39.

[3] See generally R. N. Current, *The Typewriter and the Men Who Made It* (University of Illinois, 1954).

tion and diffusion suggests that the cost of clerical labour was hardly a very pressing factor in the emergence of the new office routine.

v

There were many sectors where American manufacturers failed to make any noteworthy progress before 1860. V. S. Clark pointed to the contrast: 'In Great Britain engineering requirements demanded heavy strong tools with powerful gearing, high-speed water-cutting by broad cutters and a coarse rapid feed. Such machines made truer surfaces and saved time on large work, obviating hand finishing with files and emery.'[1] Clark was clearly arguing that the British machines were superior, not only in quality of work, but also in saving labour. Brilliant though Hoe's achievements were eventually to be in the development of printing machinery, it is significant how crude and labour-wasting was the heavy machinery he used to finish the bed plates in these early years. His planer 'was like a gigantic carpenter's plane weighted down by three or four men who stood upon it and were drawn over the iron by chains and windlasses'.[2] The effects of relative labour costs upon the development of heavy engineering in the two countries are not very obvious.

How can we explain why the Americans concentrated their energies where they did? It could be argued that they innovated most eagerly where the initial cost of the machinery, and therefore the risk, was low – in technologies requiring light, not heavy, machine tools, for example.[3] Contemporaries certainly made frequent comments about the cheapness and flimsiness of American machinery. Habakkuk's view on this is that although scarcity of labour biased American entrepreneurs towards labour-saving methods, by exerting pressure on profits, it also provided some incentive to search for ways of economizing on other factors too.[4] Within the framework

---

[1] V. S. Clark, *History of Manufactures in the United States* (Washington, 1929), Vol. I, p. 418.

[2] Ibid.

[3] W. P. Strassman, *Risk and Technological Innovation* (New York, 1959), p. 222. This line of argument is borne out by E. Mansfield, *Industrial Research and Technological Innovation* (New York, 1968), Chapter 7.

[4] H. J. Habukkuk. See p. 71 ff. of this volume – pp. 85 ff. of the original book.

of a factor proportions analysis – the relevant one if we are arguing that Americans used lighter machines of the same type as the British – such an attempt to have it both ways is theoretically unacceptable. In terms of a new technology it has to be seen in the light of our earlier discussion of this point.

Temin has suggested that American use of lighter machinery may indicate the employment of less, rather than more, capital-intensive techniques.[1] The argument has been criticized on theoretical grounds but in any case, both the Temin and Habakkuk points of view possibly founder in view of the fact that the American light, and the British heavy machines were destined for different uses. If a steam hammer was not built to last for ever, it would not work properly at all. Planers, whether made by Hoe or Whitworth, were inevitably heavy tools. Where it was technically possible, British firms turned out cheap, flimsy machines – Yorkshire became notorious for it! Habakkuk's suggestion that the building of heavy, durable equipment implies that the British had less optimistic expectations about technical progress can hardly be taken seriously in this context.[2]

We may perhaps learn more of the contrasting patterns of technological change between the two countries by looking at one sector where the lead was not uncompromisingly American – the manufacture and use of textile machinery. To 1914 textile engineering was the largest single branch of engineering in Britain: the American industry for its part, seems to have varied greatly, by sector and over time, in its response to technological opportunities.

As Habakkuk shows, for three decades after 1810, significant improvements were made to cotton textile machinery in the United States, some of which, at least, reflected a rational response to factor price differences. Both Strassman and McGouldrick state that apart from the development of ring spinning, the American industry subsequently fell into a deep

[1] P. Temin, 'Labour Scarcity and the Problem of American Industrial Efficiency in the 1850s', *Journal of Economic History*, XXVI (1966), p. 291, but see generally R. W. Fogel, 'The Specification Problem in Economic History', *Journal of Economic History*, XXVII (1967).

[2] H. J. Habakkuk. See pp. 62–3 of this volume – pp. 57–8 of the original book.

technological conservatism;[1] makers saw themselves as job order shops for individual weavers and spinners. It was the English machine builders who produced the standardized lines of equipment and ignored the whims of particular customers. Techniques, such as the large revolving flat card used in aligning the fibres and removing waste, originated in Britain and were taken up in the United States only after a long delay, even though by producing more output per operative and per machine and requiring labour of low skill, they fitted the American needs admirably. Ring spinning was first developed in the United States but it was British makers who came to dominate all world markets outside the United States, despite the slow adoption of ring spinning in the home market. The British were even more pre-eminent in worsted machinery where, in 1910, over 80 per cent of that in use in the United States came from Britain. Contrasting this with the more successful American woollen machinery industry, Habakkuk suggests that labour was less scarce when the new worsted machinery arrived some time after the woollen, and in these circumstances imported techniques were sufficiently labour-saving and required little American improvement.[2] This hardly accords with his general argument that labour-saving techniques were worth while whatever the labour costs and cheaper labour only facilitated their introduction. Strassman makes almost the opposite argument, suggesting that Americans were slow to take up worsted manufacture because of the need to comb the wool by hand first. The lower labour costs in Britain made possible mechanization of part of a process, leaving the other parts still in the hands of craftsmen.[3] Possibly this cost disadvantage left too small a market for American worsted machinery makers: this was certainly so with the very limited

[1] Davis and Stettler argue that recorded increases in productivity suggest that this view is incorrect but they are unable to provide any substantial evidence of significant technological change to disprove it. See Lance E. Davis and H. Louis Stettler, 'The New England Textile Industry 1825–60', in National Bureau of Economic Research, *Output, Employment and Productivity in the U.S. after 1800* (1966), p. 229. Productivity certainly improved as a result of the learning by doing process within the cotton factories themselves, but this does not excuse the cotton machinery makers from the charge of conservatism. See generally Strassman, op. cit., pp. 89–92 and McGouldrick, op. cit., p. 18.

[2] H. J. Habakkuk, *American and British Technology*, p. 104.

[3] Strassman, op. cit., p. 93.

American demand for spinning mules. It is perhaps significant that the large number of very small-scale weavers in Lancashire much inhibited the production of weaving machinery there too.

One factor operating to the disadvantage of American textile machine makers was that, at least until the end of the century, it was not found possible to achieve interchangeable manufacture, nor were the American machine tools well adapted to such relatively heavy work. The machining of smaller parts such as spindles, which were produced in batches of up to 10,000, did offer such opportunities, yet it appears that interchangeability was still achieved in Lancashire before 1914 by the old method of highly skilled workers working with grindstones and emery cloth. By all reports the Americans were hard pressed to match 'the superb work of the English fitters' with machine techniques.

Sandberg's article, reprinted in this volume, looks closely at a technology that was adopted much more slowly in Britain than in the United States – ring spinning. For this the Lancashire spinners have been much criticized but, as Sandberg rightly points out, the fact that they were to suffer from this delay after 1914, tells us nothing about the rationality of their behaviour before that date. In fact he tries to show that the differences in factor prices, in types of cotton used and of yarn spun and in the organizational structure of the two industries, justified the retention of mule spinning in Britain. It is a classic example of industries responding to factor price and related cost questions in the manner Habakkuk put forward.[1] Of course the latter – the fitting of a new technology into an old industrial structure – could work both ways. Some of the conservatism of American textile manufacturers derived from the difficulty of modifying their large integrated and inflexible mills. They came into their own with innovations such as the Northrop loom which required simultaneous changes in vertically related stages

---

[1] It will be seen that Sandberg actually analyses a hypothetical situation, assuming that British spinners installed mules for spinning counts higher than 40 and rings for spinning lower counts. The evidence did not allow him to show that this was in fact the case: if British spinners were installing mules for spinning lower counts, by Sandberg's analysis they were acting irrationally. In this article he only discusses the criteria for buying new equipment to expand output and not the additional criteria to be applied when old equipment was being replaced.

of production – something which, like ring spinning – fitted uneasily into the disintegrated British industry.[1]

Such a brief survey of the textile industries does at least indicate the great variety of experiences that have to be explained. Each country had its successes and failures. Sometimes there were justifiable shifts in factor proportions that the other quite properly did not follow. Sometimes the diffusion of a technology which originated in one and was relevant to the other, was surprisingly slow – the flat revolving card for example. Sometimes there were market problems, sometimes difficulties of technology. No single pattern of technological superiority emerges. Only a clear delineation of each separate problem after the manner adopted by Sandberg is likely to lead to any very fruitful conclusions.

VI

This might well be the conclusion to the whole of this survey. The experiences of British and American industries are more diverse than Habakkuk allows and the explanations therefore likely to be more varied. That the dearness of labour played a significant part in shaping American technology could scarcely be denied: that there was some straightforward shift of factor proportions is clear and unsurprising. But precisely where this occurred and, more important, if it was the major feature of the Anglo-American contrast, is not nearly so sure. The spottiness of technical achievement on both sides of the Atlantic poses a variety of problems. Why did Americans choose to substitute capital for labour in some fields and not, apparently, in others. Is it the case, as one would expect, that this was carried out most avidly in those industries where labour cost formed a high proportion of total costs? What of the price of capital and – even more important, but by no means the same thing – its availability? American railroads were constructed in a markedly less capital-intensive manner than those in Britain at the outset and the reason was much more the inability of the promoters to accumulate sufficient capital, rather than technological or relative factor price considerations. Possibly at a later stage they were able to modify the technology,

[1] See M. T. Copeland, *The Cotton Manufacturing Industry of the United States* (Cambridge, Mass., 1912), pp. 87–92.

though they would then come up against the problem of related costs. In any case this only goes to show that great care
must be taken to identify the stage of development of the industries being compared in the two countries.

How far is it true that British makers held on to labour-intensive techniques, in what sectors, and with what justification?
This is a huge question but we cannot proffer general explanations until we are in a position to answer it with some degree
of precision. A careful distinction must be drawn between
invention, innovation, and diffusion. There is little evidence
that the Americans were more inventive than others and indeed
it is remarkable how many inventions originating in Europe –
in the steel industry for example – were of particular relevance
to American operating conditions.[1] We know very little indeed
about diffusion, yet the issue may be a critical one, for Temin
hazarded the opinion that maybe the main element in the Anglo-
American contrast was an asymmetry in the flow of knowledge,
that American manufacturers were more interested in learning
from Europe than vice versa.[2] In the article reprinted in this

---

[1] McCloskey has argued that British steel makers did not take up new American
technologies after 1880 because they were not relevant to the British cost position.
In terms of total productivity, although the Americans caught up the British
before 1914, they did not surpass them. D. N. McCloskey, 'Productivity Change
in British Pig Iron, 1870–1939', *Quarterly Journal of Economics* LXXXIII (1968).
Of course the steel industry in Britain might still be lagging behind average
productivity in growth in British industry generally and therefore be at an
increasing disadvantage in international trade. It is curious that McCloskey
takes no account of by-product recovery which was one area where Britain was
said to have fallen behind seriously and his type of analysis asks only why
Britain did not *follow* American practice, not why they ceased to push out the
frontiers of technology for themselves.

The contrast between two analyses of the same situation is interesting. Temin
has argued that British manufacturers did not have the opportunity to install the
latest plant because demand was increasing so slowly. (P. Temin, 'The Relative
Decline of the British Steel Industry, 1880–1914' in H. Rosovsky, *Industrialization
in Two Systems*, New York, 1966.) McCloskey's point is that it was not profitable
for the British to introduce American methods even when new capacity was
required.

[2] P. Temin, 'Labor Scarcity', p. 295. Easterlin, noting the relatively high level
of formal education in the U.S.A., Germany, and Sweden in the nineteenth century
suggests that 'it is unlikely that relative factor prices in these countries differed
in a way uniformly encouraging the modification of British technology', and
suggests that a great deal depended on the way in which each country learned
the new techniques. R. Easterlin, 'A Note on the Evidence of History', in C. A.
Anderson and Mary Jean Bowman (eds), *Education in Economic Development*
(Chicago, Ill., 1965), p. 423.

volume, I voice the suspicion that some critics compare average British practice with best practice elsewhere and rarely seek to complete the investigation. The decentralized and small-scale operations of much of British industry probably meant that there was an unusually large gap between best and average, though it is by no means certain. In engineering, technological change was not just a matter of buying machines but possibly even more one of workshop organization and control. Ideas of this kind took a long time to percolate through and there Britain depended heavily on American managers who came to teach by doing.

Statistical investigations have established that, in general, there is a link between the size of a firm and the speed of its response to innovations. The bigger they are the more likely it is that, at any one time, there will be units of equipment to replace. With a wider range of operating conditions, large firms have a better chance of containing the optimum technical relationships for the introduction of an innovation. No thorough historical investigation has been made of relative sizes of firms and plants, but such meagre evidence as there is suggests that the largest plants in the engineering and metal trades in the United States were bigger than those in Britain by a considerable margin. Furthermore, British industry contained a mass of small jobbing shops where significant technical innovations could be introduced only with great difficulty. My article seeks to show that the market was a critical factor in impeding both product innovation and process changes. The problem was not the size of the total market but particular factors impeding the development of individual industries. Either demand grew too slowly to warrant the employment of mass production techniques, or the structure of demand was unfavourable – with locomotives for example. Yet where demand conditions were more favourable, American manufacturing techniques were followed with great success – bicycles and high-speed steam engines, for example.

The unique element in American technology was the development of the machine tools required for mass interchangeable manufacture. The elements leading up to this were many – a purely technical solution to an engineering problem, a pattern of demand favouring the standardized goods produced by such techniques, labour costs, low initial cost of the machines, absence of older and different engineering traditions, and so forth. Whatever the driving forces, the technological fall-out to American industry at large from this advanced machine tool sector was very substantial. Witness, for example, the trained machine tool engineers who played a vital part in the development of the United States automobile industry. Technological progress was self-reinforcing in another sense too. The degree of specialization and of size achieved by the American machine tool industry was possible only because of the simultaneous growth of machine-using firms enjoying certain technological processes in common. Habakkuk rightly suggests that in Britain technological progress more often took a form which was not capable of wide application outside the industries in which it originated. The practice of many firms of making their own machinery is relevant here too.

VII

It has not been possible to do more here than survey some of the many vital issues raised by Habakkuk's book. This discussion has been confined to those factors directly affecting choice of techniques, but in so far as comparative rates of growth of total demand influenced the adoption of technologies in the two countries after the Civil War, the question broadens into an investigation of the factors underlying the slow growth of the British economy as a whole. This issue is taken up at some length by Habakkuk but is one that we have not followed up in these readings.

As we have seen, Habakkuk's explanations of the observed technological divergencies are couched either in terms of rational responses to different factor prices, or of American adoption of techniques – either new technologies or shifts in factor proportions that saved both labour and capital – which would have been profitable in Britain too but for some reason were not taken up. One serious difficulty is that we do not in many cases know that American methods produced goods at prices that were competitive with those of British manu-facturers. The tariff and transport costs could give American makers an advantage that was as high as 45 per cent in the price of textile machinery in the American market, for example.

Where American manufacturers competed with British on equal terms it was, as often as not, new products that they were offering. Then there arises the problem of choosing between these economic and the non-economic explanations. Here Habakkuk is not always very helpful, because although in the end he comes down on the side of the former, his book abounds with references to the latter, even to the point of describing the adoption of labour-saving devices in the United States as a 'habit'. The timing of the emergence of the unique system of American manufacture is another question. It is central to Habakkuk's thesis that the method was established by 1850, before the inflow of immigrants came to alter the labour supply position entirely. In this introduction doubts are cast as to the extent of American technological advance by that time, and Temin and others have pointed to significant areas where labour-saving techniques were most conspicuous for their absence during those years. This, combined with the still unresolved problem of the precise determinants of American real wage levels at that time, leaves a wide area for research and speculation.

Despite all these difficulties, however, the true merit of Habakkuk's book lies in the challenge it offers. Ranging so widely and courageously over difficult theoretical issues and only partly researched historical situations, it is sometimes hard to follow and its conclusions are not always easy to accept. But it forces us to look squarely at one of the most important issues of recent times and to put aside a parochial attitude towards history. One can only hope that it will encourage research, not only into this comparison, but also into the possibly more rewarding field of Anglo-European technologies.

1 The Economic Effects of Labour Scarcity

H. J. HABAKKUK

[This material is extracted from *American and British Technology in the Nineteenth Century* by H. J. Habakkuk (Cambridge University Press, 1962).]

LABOUR-SCARCITY AND CHOICE OF TECHNIQUE

The supply of labour in the U.S.A. and England

Industry started to develop in the United States at a time when industrial money wages were substantially higher than they were in England, according to some estimates for the early nineteenth century perhaps a third to a half higher.[1] This was fundamentally because the remuneration of American industrial labour was measured by the rewards and advantages of independent agriculture.[2]

[1] In the colonial period American 'workmen commanded wages from 30 to 100 per cent higher than the wages of contemporary English labouring men'. (*The Growth of the American Economy*, ed. H. F. Williamson, pp. 49–50, 53, 101, 137.) According to Nassau Senior, writing in 1829, labourers' wages in North America were 'twenty-five per cent higher than they are in England, while the labour requisite to obtain necessaries is not much more than half as great in the former country as in the latter'. (*On the Cost of Obtaining Money* (London, 1830), pp. 2, 11, London School of Economics Reprints No. 5.) ' That the general rate of wages is higher in the United States than in Britain is admitted, particularly the wages of females employed in the Factories.' James Montgomery, *A Practical Detail of the Cotton Manufacture of the United States . . . Compared with that of Great Britain* (Glasgow, 1840), p. 135.

[2] V. S. Clark, *History of Manufactures in the United States* (Washington, 1929), Vol. I, pp. 152–8. The clearest statement of this connection was put by Hamilton in the process of arguing against it. 'The smallness of their population compared with their territory; the constant allurements to emigration from the settled to the unsettled parts of the country; the facility with which the less independent condition of an artisan can be exchanged for the more independent condition of a farmer; these and similar causes, conspire to produce, and, for a length of time, must continue to occasion, a scarcity of hands for manufacturing occupation, and dearness of labour generally.' (A. Hamilton, *Report on Manufactures*, p. 123.) 'The price of manual labour, and the compensation of what is purely mechanical, such as the salary of ordinary clerks, is much higher than in Europe. This is a simple result of the comparative state of demand and supply for those objects, arising principally from the superabundance of land compared with the present

B

Where land is in such plenty [wrote an observer in the 1760s] men very soon become farmers, however low they set out in life. Where this is the case, it must at once be evident that the price of labour must be very dear; nothing but a high price will induce men to labour at all, and at the same time it presently puts a conclusion to it by enabling them to take a piece of waste land. By day-labourers, which are not common in the colonies, one shilling will do as much in England as half-a-crown in New England.[1]

Land was abundant and, except possibly in Virginia before the abolition of entails, it was accessible to purchase by men of small means. In the 1820s it could be purchased from the Federal Government at $1.25 per acre, which was well within the reach of a skilled labourer, who might at this time earn between

---

population.' Gallatin to La Fayette, 12 May 1833, *Writings of Gallatin*, ed. Henry Adams (Philadelphia, 1879), II, p. 471. To argue that, in the first half of the nineteenth century, agriculture set a floor to American industrial wages, it is not necessary to suppose that large numbers of industrial workers moved west. It is sufficient if 'the abundance of western land drew away many thousands of *potential* wage earners (from the hill towns of New England and from the exhausted farms of New York and Pennsylvania) who might otherwise have crowded into the factories'. (C. Goodrich and S. Davidson, 'The Wage-earner in the Westward Movement', *Political Science Quarterly*, LI [1936], p. 115.) Thus the argument does not depend upon Turner's frontier hypothesis, rigorously interpreted. A large amount of work has been devoted to demolishing this thesis; the work that is most relevant to our argument attempts to show (*a*) that the 'West' lured more Europeans to the Eastern states than Easterners to the West, and (*b*) that there was a net migration from farming families to the towns. (See F. A. Shannon, 'A Post-Mortem on the Labour-Safety-Valve Theory', *Agricultural History* (1945). The demonstration has been made for the period *after* 1860 and therefore does not bear directly on our argument. Moreover, even for the period after 1860, Professor Shannon's conclusions are likely to underestimate the effect of returns in agriculture on wages. He compared the actual farming population in 1900 (28 million) with his estimate (40 million) of the population which would have been produced by the natural increase of the farming population of 1860 (19 million), and he concluded that the excess of 40 millions over 28 millions represented migration out of agriculture. But his estimate of 40 millions was arrived at by multiplying the farming population of 1860 by an 'average rate of breeding' which was not, in fact, the rate of natural increase but *included* increase by immigration. Moreover, in his calculations, he restricted the frontier to the 'farming frontier'; but from the point of view of our argument the 'pull' of the frontier towns, which develop because of the movement of the farming frontier, is also relevant. On this subject I am indebted to William Fredrickson.

[1] *American Husbandry*, ed. H. J. Carman (New York, 1939), pp. 53–4, quoted by D. L. Kemmerer, 'The Changing Pattern of American Economic Development', *Journal of Economic History*, XVI, 1956, p. 576.

$1.25 and $2.0 per day.[1] 'The men earn here (in the cotton-textile factories at Lowell) from 10 to 20 dollars a week,' wrote an English observer in 1842, 'and can therefore lay by from 5 to 10 dollars, after providing for every want, so that in two or three years they accumulate enough to go off to the west and buy an estate at 1¼ dollars an acre or set up in some small way of business at home.'[2] In England, by contrast, land was scarce in relation to labour, and the supply of land on the market, particularly of small properties, was restricted by the existence of large estates supported by legal restraints on alienation; the return on the cultivation of land in England up to 1815 was high since new techniques were available, and food prices were rising; but to set up as a tenant-farmer required considerable capital, and, even if an English artisan had been able to acquire the capital, the supply of farms to be let was limited, and absorbed by the demand of the sons of existing tenants. In England, therefore, a man could, generally speaking, enter agriculture only as a labourer commanding low wages.

The abundance and accessibility of land plus the fact that much of it was fertile meant that output per man in American agriculture was high. Moreover, since the cultivator was often also the owner, and his family supplied the labour, the advantage of the high output accrued to the cultivator. His income included: (1) an element of rent, which would in England have been a heavy charge on output and payable to the landowner, (2) agricultural profits, which in England accrued to the tenant-farmer, as well as (3) the wages, which in England went to the agricultural labourer. Furthermore the new land was brought into use in such a way that the returns to settlement on the frontier sometimes included elements of exceptional gain. Many American farmers had heavy debt charges; but there

[1] It was available at this price after 1820 if bought in tracts of 80 acres (the units previously laid down were 640 acres (1785), 320 acres (1800), 160 acres (1804). There were in addition the costs of bringing it under cultivation. 'Buying the land is nothing; the real expense is getting it ready, on account of the cost of labour.' In Michigan in 1831, one of de Tocqueville's informants estimated that to get the land under cultivation cost $5 to $10 per acre. Another settler gave a lower estimate: to establish a new settlement one needed between $150 and $200, 100 of which was to buy the 80 acres and the rest for the expenses of initial settlement and incidentals. (de Tocqueville, *Journeys to England and Ireland*, ed. J. P. Mayer (1958), pp. 209-10, see also p. 343.

[2] J. S. Buckingham, *The Eastern and Western States of America* (1842), I, pp. 300-1.

were no tithes, and taxes were low. Finally since the total earnings of the family in American agriculture tended to be divided among its members, there was less disparity between average and marginal earnings. In order to attract labour, therefore, industry had to assure the workers in industry a real wage comparable to average earnings in agriculture. English industry, by contrast, could acquire labour from agriculture at a wage equal to the very low product of the marginal agricultural labourer plus an addition to cover the costs of transport and of overcoming inertia. Thus, while in England industrial wages equalled the marginal product, in the U.S.A. the reward of the marginal labourer in industry was above his product, unless the manufacturer took steps to increase the product.

Moreover the course of agricultural technology in the early decades of the nineteenth century may well have accentuated the disparity between the terms on which labour was available to industry in the U.S.A. and England. In America improvements in agriculture took the form primarily of increasing output per head and the increase initially was probably more rapid than in industry; in England on the other hand, agricultural improvement was devoted primarily to increasing yields per acre and, even where there was an increase in output per head, the abundance of labour made it difficult for the labourer to enjoy the increase. In America agricultural improvements raised, and in England prevented, a rise in the terms on which labour was available to industry.

Comparison of industrial wage-rates does not of course measure precisely the difference in labour-cost, in terms of output, in the two countries. For, on the one side, the hours of work in America were generally longer,[1] and conceivably effort was more concentrated. It may also be, as some contemporaries said, that better nutrition and more spacious working and living conditions, made the American a more efficient worker. On the other hand it might be supposed from the alternatives open to labour in America that the workers recruited into American industry were inferior, in relation to agricultural labour, when compared with the English; it is possible that is, that the pull of

[1] 'The number of hours constituting a day's work especially in factory labour, is much greater; and . . . this excess in the hours of work obtains generally in every industrial occupation.' (P.P. 1854, XXXVI, p. 14.)

agriculture showed itself in the quality as well as the price of industrial labour in America. Then again, American employers had to incur higher costs than the English on housing and working-facilities of a kind which made the work more agreeable but added little, if anything, to labour productivity. Probably also the rate of turnover of workers was higher in America, and therefore the likelihood smaller that they would acquire industrial discipline. Finally, England at the start was technically superior and this must have reflected itself in higher labour productivity. Since we cannot measure these conflicting influences, it is impossible to be precise about the differences in labour-costs in the two countries, but there is no reason to doubt the opinion of contemporaries that American industrial labour was substantially dearer than the English.

But American industrial labour was not only dearer than the English; its supply was less elastic.[1] It was more difficult to

---

[1] M. D. Morris ('The Recruitment of an industrial Labor Force in India, with British and American Comparisons', *Comparative Studies in Society and History*, II (1960), pp. 305–28) argues that in neither the U.S.A. nor Britain did the recruitment of labour significantly inhibit industrial expansion in the cotton-textile industry, but concedes that in the U.S.A. 'during the periods when the industry was expanding rapidly the new labor force did not always automatically flow in as rapidly as was necessary to fill the mills with labor. In this sense, the situation was different (from the) typical British case' (p. 319). It is impossible to say how great a disparity in this respect existed between the two countries, especially since there were considerable regional variations within the U.S.A. itself. But that there was some disparity seems a fair deduction from contemporary comment. 'Employment (in America) . . . is always plentiful: labour, especially skilled labour, ever in demand.' 'Full employment can always be obtained by competent workmen.' (P.P. 1854, XXXVI, pp. 14, 46.) The American manufacturer 'must often carry on his business at little or no profit, perhaps at a considerable loss, in order to keep together the agents by which, when the demand comes, he can alone supply it. Hence the anxiety to settle down the operatives around the mills; to render their condition and social position such as shall absolutely attach them to their employers and the locality. The latter, perhaps, being the most difficult, from the migratory tendencies of a people so restless, and always so alive to any new contingency which promises to better their condition, however distant the field of operation may be.' (P.P. 1854, XXXVI, p. 42.) It was particularly difficult to get together enough skilled operatives after a firm had reduced its operations, because they sought and obtained employment in other districts. '. . . in Boston a constant deficiency of labour had seriously hampered the growth of industry until the forties. . . . Few (industries) grew significantly (1835–1845), and many actually declined. The prospective manufacturer desiring a site for a new establishment, or the capitalist with an "abundance of money seeking an outlet" found little encouragement. And even those already established who wished to expand were inhibited by the apparently inflexible labour supply.' (O. Handlin, *Boston's Immigrants* (2nd ed. Cambridge, Mass. 1959), p. 74.)

obtain additional labour for the industrial sector as a whole. This was partly owing to the abundance of land and the difficulties of internal transport, which required technical solutions and heavy capital outlays before they could be overcome. It was also partly owing to America's geographical remoteness from the areas of abundant population. The U.S.A. in the early decades of the nineteenth century was not only sparsely populated, but was forced by the high costs of transport (including the costs of overcoming inertia) to depend for labour, down to the 1840s, mainly on her own resources. And though the American rate of natural increase of population was exceptionally high, a large part of it was absorbed in settling new land, and the labour to work additions to industrial capacity tended to be difficult to obtain. The Superintendent of the Springfield Armoury wrote in 1822: 'I find the rage for manufacturing cotton prevails to such a degree and there is so great a call for first-rate workmen that I am apprehensive I shall lose some of our most valuable workmen except I am authorised to raise their wages according to circumstances.'[1] By contrast, the industrial areas of England were near to densely populated agrarian slums in the southern agricultural counties and in Ireland. For an identical rate of increase of capacity, therefore, the American industry would have had more difficulty than the English in obtaining additional labour.

Dearness and inelasticity are logically distinct; the consequences of a high level of wages prevailing over time are not the same at all points as those of a wage-level which rises when demand for labour increases. There have been situations where the floor set to industrial wages by *per capita* productivity in agriculture was low but the supply of labour was inelastic, for example in England for much of the eighteenth century. There have also been situations where the floor set to industrial wages was high but where abundant labour was forthcoming at this wage – in some respects this was the case in the U.S.A. in the second half of the nineteenth century when there was a large amount of immigration. But in the first half of the nineteenth century, with which we are here mainly concerned, there was a contrast on both points, and a more marked contrast than

[1] F. J. Deyrup, *Arms Makers of the Connecticut Valley* (Smith College Studies in History, Vol. XXXIII; Northampton, Mass. 1948), p. 105.

existed either before or since; industrial labour was dear, and its total supply inelastic in the U.S.A., and it was cheap and elastic in England. Since the general level of labour-costs was so closely connected with the elasticity of the supply of labour, it is difficult to discuss their effects separately except at the cost of repetition. In most of the following discussion, therefore, they are treated together.

The inducement to mechanize

It seems obvious – it certainly seemed so to contemporaries – that the dearness and inelasticity of American, compared with British labour, gave the American entrepreneur with a given capital a greater inducement than his British counterpart to replace labour by machines. The real problem is to determine how the substitution took place. Where the more mechanized method saved *both* labour and capital per unit of output it would be the preferred technique in both countries. It was where the more mechanized method saved labour but at the expense of an increase in capital that the American had the greater inducement than the English manufacturer to adopt it. [The term capital-intensive will henceforward be used to describe such a method.]

A number of situations can be distinguished in which this inducement to make a more capital-intensive choice of technique would operate.

(*a*) Provided final product-prices did not rise – or, at any rate, did not rise in the same proportion – a given rise in wages in terms of output reduced the rate of profit in the method with a low output per man but low capital-intensity more than in the method which had a higher output per man because it was more capital-intensive, *even if* the price of machines rose in the same proportion as wages. If the price of machines did *not* rise in the same proportion as wages, the relative advantage of the capital-intensive method would, of course, be even more pronounced. Given the assumption about product-prices, this is simply a matter of arithmetic.[1] The effect of a rise in wages in a situation broadly analogous to the one just described was first

---

[1] J. Robinson, *The Accumulation of Capital* (1958), p. 107. D. H. Robertson, *Lectures on Economic Principles* (1958), pp. 110–13. In this section I have drawn heavily on Mrs Robinson's book.

analysed by Ricardo, and in his model product-prices remained constant (as well incidentally as machine-prices) because, in his view, the general level of prices was determined by the gold supply.[1] If one rejects Ricardo's view of the way in which the general price level is determined, it is not easy to see why, in a closed system and the long run, the price of products, like the price of machines, should not rise in the same proportion as wages, in which case there would be no inducement to shift towards more capital-intensive methods. But, in real life, it would take time before the rise in wages worked itself out, and the price of final products and also of machines would lag behind. For some time, most final products would be produced by the existing stock of machines, made with labour before the rise in wages, and the same would be true of new machines. In a world where price was equal to marginal cost this would mean only that manufacturers who produced with the existing stock of machines would enjoy a period of quasi-rents. But in practice it is unlikely that prices would rise to this extent. Moreover we are not considering a closed system, and it is reasonable to assume, for the American economy in the first half of the nineteenth century, that product-prices did not uniformly rise in proportion to the cost of labour in the case of those products in which there was an international price to which American prices tended over a period to accommodate themselves; that is, when imports prevented the American producer passing on the rise in his labour-costs to the consumer.

Moreover, a similar bias towards the most capital-intensive methods would be exerted if, when product-prices fell – because of a deflation of demand – money wage rates did not fall proportionately; and it is probable that the great alternatives to industrial employment in the U.S.A., while they compelled the manufacturer to pay much higher wages for additional labour, also made it difficult, in the short run, to reduce wage rates when product-prices fell, for one consequence of

[1] D. Ricardo, *Works and Correspondence*, ed. P. Sraffa (Cambridge, 1951), I, ch. XXXI, p. 395. 'The consequence of a rise in food will be a rise in wages and every rise of wages will have a tendency to determine the saved capital in a greater proportion than before to the employment of machinery. Machinery and labour are in constant competition, and the former can frequently not be employed till labour rises.'

labour-shortage in America was that labour was often hired by yearly contract.[1]

(*b*) The argument in the preceding paragraphs has assumed that the construction-time of machines was so short that it can be ignored, and we have not considered the effect on the relative cost of different techniques of the rate of interest in the calculation of capital-costs. We shall now consider what happens when this assumption is dropped. The cost of capital can be considered as consisting of two parts, the principal, that is the price of the plant, and the rate of interest which the manufacturer applied to this principal. If the principal is thought of as consisting solely of labour, it would rise in the same proportion as the rise in the price of labour. But, on Ricardo's assumptions, the total interest bill would not rise in the same proportion, because the rate of interest applied to the capital would now be lower. If manufacturers, in calculating the cost of capital goods, employed a notional rate of interest equal to the rate of profit, this would in general tend to increase the relative attractions of the more capital-intensive technique, because capital would be cheapened relatively to labour.

It is, however, important to notice that there are certain circumstances in which, on our present assumptions, a rise in wages might have the reverse effect, that is, might induce the adoption of a technique which was less capital-intensive and less productive per unit of labour, though still the most profitable in the circumstances.[2] And since this possibility is of more than theoretical interest it needs to be explored. Capital-cost consists of the outlays on new plant plus the interest on these outlays during the construction-period of the plant minus the interest

[1] 'The greater part of the (cotton textile) Factory workers being connected with farming, whenever wages become reduced so low as to cease to operate as an inducement to prefer Factory labour above any other to which they can turn their attention, then a great many Factories will have to shut up. During a stagnation in trade, it is common for the manufacturers here (i.e. in the U.S.A.) to stop a part, or the whole of their Factories, and then the workers retire to their farms; such was the case in 1837, when a vast number of Factories were entirely shut up.' (J. Montgomery, *The Cotton Manufacture of the United States Contrasted and Compared with that of Great Britain* (Glasgow, 1840), p. 137.)

[2] See Joan Robinson, op. cit., pp. 109–10, and C. A. Blyth, 'Towards a More General Theory of Capital', *Economica*, XXVII, May 1960, pp. 126–8.

on the amortization funds out of the earnings of the plant during its lifetime. At the higher level of wages, the rate of profit and therefore, on the present assumptions, the rate of interest employed in calculating capital-costs, would be lower than before. There are circumstances in which, at this lower rate of interest, the less capital-intensive technique will be preferable; though its output per man is less, its capital-costs are now lower in a greater degree. When the capital-intensive technique has a much longer construction-period, in relation to its lifetime, than the labour-intensive technique, that is when it locks up capital for a long period before any output is forthcoming, the entrepreneur at the higher level of wages may find it more profitable to employ the technique which has the shorter construction-period in relation to its lifetime, that is, which starts to yield its output sooner, even though the output per unit of current labour is lower.

The circumstances in which this may occur have been analysed. The first condition is that the choice lies between techniques which vary considerably in the ratio between their construction-time and their lifetime, the construction-period being longer, in relation to lifetime, in the capital-intensive than in the labour-intensive technique. The second condition is that the relation between construction-time and lifetime is fixed within narrow limits by technical considerations, so that the techniques available do not form a spectrum and the entrepreneur has to choose between radically different ways of using his resources. But how frequently are these circumstances likely to have occurred in practice? The theoretical case has been illustrated by a comparison between the production of meat by the fattening of cattle and by the grazing of sheep, the first standing for a process which yields its product only after a long period of expensive preparation, but then yields a very large product for the labour employed, the second a process which yields a small output per man, but yields it sooner and more continuously.[1] At a high level of wages, the balance of advantage may turn against the more capital-intensive labour-productive method which locks up its capital for a longer period. Agriculture might provide other instances, for shifts from one crop to another, or from arable to pasture, necessitate radical changes in organization. There may sometimes be analogous choices in mining or in the production of primary products where the plantation is an alternative to peasant agriculture.

A rise in wages *may*, then, provide an incentive to adopt techniques which are less capital-intensive. But this effect is a possibility only where a longer construction-time is linked, by technical necessity, with a shorter lifetime. And it is difficult to think of examples in manufacturing proper. In manufacturing, there is more likely to be a continuous range of known techniques, such that the entrepreneur can move smoothly along the range, making small adaptations all the time; in so far as there are differences in this respect, it is generally the more capital-intensive and labour productive of the known methods which are the most durable, that is, have the longer lifetime in relation to their construction-period. While therefore it is not a universal rule that a rise in wages will change the ranking of techniques according to profitability in favour of those which are both more productive per unit of labour and more capital-intensive, it is justifiable to suppose that this is the common case in industry.

(*c*) In paragraphs (*a*) and (*b*) above we have been concerned with variations of the 'Ricardo effect'. But quite apart from this effect, a rise in wages would have raised the cost of capital-intensive techniques less than that of labour-intensive techniques if it was easier to import machines than labour from countries where labour was cheaper and its supply more elastic. Even if wages in America rose in the same degree for all types of labour, the attractions of the more capital-intensive methods would still have become relatively greater to American manufacturers, as a result of the rise in their labour-costs, if they could import capital-intensive equipment from areas where labour-costs had not risen.

(*d*) There may also have been significant differences in the elasticity of supply of different *types* of labour and in the extent to which capital- and labour-intensive techniques used the different types; and where the more capital-intensive techniques made relatively greater use of the type of labour which was most abundant, a general rise in wages would favour their use.

There is little readily available information about the labour requirements of various techniques or about the price of different types of labour, and the following discussion is therefore conjectural. But a plausible case can be made for supposing that in the early nineteenth century in the U.S.A. an increased demand for labour raised the wages of skilled labour *less* than the wages of unskilled labour, and that, in many cases, the capital-intensive technique required, for its construction plus operation, more skilled labour per unit of output than the labour-intensive technique.

Was the differential for skill smaller in America than in England? V. S. Clark, referring apparently to the 1820s, considered that differences in wages between England and America were greater in unskilled than in skilled occupations.[1] On the other hand some incidental comment suggests that it was difficult to train a class of skilled workers in America at least in textiles. 'During the first half of the (nineteenth) century, when the ring-frame was being introduced and when the operatives were native-born, the labour-force in the mills was constantly changing . . . so that no skilled class was developed.'[2] And Whitney described the leading object of the system of interchangeable parts as 'to substitute correct and effective operations of machinery for that skill of the artist which is acquired only by long practice and experience; a species of skill which is not possessed in this country to any considerable extent'.[3]

The distinction between skilled and unskilled labour is, of course, constantly shifting; technical progress creates new categories of employment and calls for continual redefinition of skill. Skilled labour at the beginning of the nineteenth century was very different from skilled labour at the end. At the end of the century there was a whole spectrum of degrees of skill. At the beginning of the century there were three broad categories of labour. First there was the undifferentiated mass of unskilled

[1] V. S. Clark, op. cit., I, p. 392. For further discussion of this point see pp. 128–31, 151–6, in H. J. Habakkuk, *American and British Technology in the Nineteenth Century* (Cambridge, 1962).

[2] M. T. Copeland, *The Cotton Manufacturing Industry of the United States* (Cambridge, Mass. 1912), p. 73.

[3] Quoted in J. W. Roe, *English and American Tool Builders* (New Haven, 1916), pp. 132–3.

adult labour. The money wage of such labour was a third or a half higher in America than in England. Secondly there were workers who performed tasks which required dexterity and aptitude but which, granted these qualities, could be performed after a short period of training and experience, for example some of the tasks performed by women in the textile industry. For such labour the American wage was rarely more than 20 per cent higher than the English.[1] Finally there were the craft skills which were so technically complicated that they could be acquired only after a long term of training. Craft operations were so diverse that, more than in other types of labour, it is extremely difficult to make direct comparison, especially as rates varied widely according to season and from place to place, and we have no independent tests of the degree of skill being priced. Only a very detailed analysis of labour capabilities and of the relative values placed upon them in the two countries would establish the differences in this respect between America and England. But the random selection of rates given by Clark suggests that the premium on artisan skills was generally lower in America than in England in the early nineteenth century.

How far differences in the premium on skill between the two countries represent difference in supply and how far differences in demand it is impossible to say, but there are some general reasons why we should expect the supply of skill, in relation to common labour, to have been more abundant in America than in England:

(1) As has been evident since 1939, a general shortage of labour is most acutely felt in the unskilled grades, in a shortage of recruits for heavy tedious work; workers in low-paid activities are more prepared to leave their jobs and seek better ones when labour is scarce, in relation to demand, than when it is abundant. A general shortage of labour raises the labour-costs of instrument-users more than those of instrument-makers. Where there is a persistent surplus of labour, it is those who are without skill, particularly the newcomers to the labour market, who have most difficulty in finding work; it is on the wages of the unskilled that the surplus has most effect.

[1] V. S. Clark, op. cit., p. 397.

(2) The pulling or retaining power of American agricultural expansion was felt most on unskilled labour. It was, of course, easier for the skilled worker to accumulate the capital necessary for settlement; but at the opening of the nineteenth century the costs of settlement were probably sufficiently low in relation to industrial earnings not to restrict the possibility to the highest-paid workers, and it was the worker without special industrial skills who stood to make the largest relative gain from agriculture. Furthermore, investment in social overhead capital, particularly transport-systems, made heavy demands on general labour, that is labour not trained for particular operations, and the construction of canals, roads, and railways seems to have been more attractive to such labour than the factories. This type of investment was a more rapidly increasing proportion of total investment than in England.

(3) Literacy was more widely diffused in America and popular education developed earlier. 'There are very few really *ignorant* men in America of native growth'[1] wrote Cobbett. Thus a higher proportion of the population than in England was capable of being trained to skilled operations.

(4) There was much more international mobility of skill than of general labour, and a high proportion of English migrants to the U.S.A. before the start of mass migration were skilled workers. In the early decades of the century therefore, immigration did more to alleviate the shortages of artisan skills than of unskilled labour.

(5) Mechanical abilities of a rudimentary sort were widely spread in the U.S.A. at the opening of the century. 'The manufacturing enterprises which existed in the heart of America's eighteenth-century mercantile-agricultural economy were numerous and diversified . . . varied and dextrous mechanical abilities were all but universal.'[2]

(6) The up-grading of unskilled to skilled labour was less impeded in the U.S.A. because, though skilled workers were

---

[1] Cobbett, op. cit., p. 197.

[2] G. S. Gibb, *The Saco-Lowell Shops: Textile Machinery Building in New England 1813-1949* (Cambridge, Mass. 1950), p. 10.

in fact organized earlier in the U.S.A., trade union restrictions, conventions, apprenticeship rules, were less effective than in England.

We may reasonably conclude, therefore, that in America an increase in the demand for labour raised the cost of the methods which required a great deal of unskilled labour more than it raised the cost of the methods which required a great deal of skilled labour.

It is not always the case that the capital-intensive methods require more skilled labour per unit of output than the labour-intensive. But in the technology of the early nineteenth century there are likely to have been several cases where it did so. The *manufacture* of power-looms required more skill than the manufacture of hand-looms; and the same was probably true of the 'superior' as compared with the 'simpler' machines of all kinds. In the U.S.A. when demand for labour rose, the labour-costs of the machine-makers rose less than the labour-costs of the machine-users, and the costs of the machines which were expensive in terms of output rose less than the cost of the cheaper machines. At the very start, of course, 'power-looms' may have been more expensive in terms of 'hand-looms' in America than in Britain, because American deficiencies in engineering skill, compared with the British, were more marked in the making of complicated than of simple machines. The point is that, with the increase in industrial capacity, the ratio between the costs of manufacturing the equipment fell more rapidly in America than in Britain, quite apart from any possibility that the Americans were catching up on the English in engineering skills.

The position of the *operating costs* of capital-intensive and labour-intensive techniques is not so clear. There may have been industries in which the operation of the capital-intensive machine required a *lower* ratio of skilled to unskilled labour, which, to a greater or lesser degree, offset the higher ratio in the costs of its manufacture. But the probability is that, in a significant number of cases, the manufacture plus use of the more capital-intensive techniques required more skilled to unskilled labour than the labour-intensive. Where this was so, the fact that unskilled labour, in relation to skilled labour, was dearer in America than in England gave the American an inducement to make a more capital-intensive choice of technique. There is also the additional point that the type of labour which was relatively dearest performed the simple, unskilled operations which were, from a technical point of view, most easily mechanized.

Thus there were at least four circumstances in which a rise in the cost of American labour provided the American manufacturer with an incentive to adopt the more capital-intensive of known techniques, in order partly to check the rise in wages and partly to compensate for it. As, from experience, it became evident that labour was the scarcity most likely to emerge during a general attempt to expand capacity, we should expect an increasing number of American manufacturers to have this in mind when choosing equipment and to become conditioned to adopting the method which did most to alleviate this particular scarcity. It is not, however, strictly necessary to assume that all, or indeed any, manufacturers consciously reflected in this way on the resource-saving characteristics of different techniques. Investment may have adapted itself to relative factor-scarcities by a process of natural as well as of conscious selection. If some manufacturers, for whatever reason, adopted improvements which were more appropriate to the factor-endowment of the economy, these men fared better than those which made a contrary choice. They competed more successfully in product- and factor-markets and by expanding their operations came in time to constitute a larger share of their industry.[1] Moreover they brought influence to bear not only via the market but by force of example. Their success inspired imitators, and shaped entrepreneurial attitudes towards the most likely lines of development.

In England, where the supply of labour to industry as a whole was elastic, there was no reason, so far as labour-supplies were concerned, why accumulation should not proceed by the multiplication of machines of the existing type. Thus even if the general level of labour-costs had not been higher in America, the difficulty of attracting additional labour might have pushed the American entrepreneur over a gap in the range of techniques and induced him to adopt one which was not only more

[1] W. Fellner, *Trends and Cycles in Economic Activity* (New York, 1956), p. 50.

capital-intensive than he had previously employed, but was also more capital-intensive than those currently adopted by his English counterpart. But the fact that American labour was also dearer than the English provided the American entrepreneur with an incentive to adopt more capital-intensive techniques than the English, even had the supply of labour been equally forthcoming.

For this argument it does not seem essential that the cost of finance in industry should have been lower, in relation to labour-costs, in the U.S.A. than in Britain. It is enough if machines were cheaper in America in relation to labour, either because they could be imported or because they were made with the type of labour which was relatively most abundant. But the bias towards capital-intensity would obviously have been greater if in fact finance was cheaper in relation to labour in America.

The cost of finance to a manufacturer who reinvested his profits is more ambiguous than the cost of labour, because we do not know what, if any, imputed rate of interest was used. It follows from the assumptions of the simplified version of events to which we have previously referred that the country with the higher wages in terms of output has the lower rate of profit on capital, and this can be regarded as, in some sense, the relevant rate of interest. But this is not a very helpful guide to what actually happened. For though there are many quotations of profits for particular firms in particular years, it is rarely clear what definition has been employed and it is difficult to derive any general rate of profit.[1] In any case accounting methods in early nineteenth-century manufacturing were extremely rudimentary, and it is wildly unlikely that, in comparing alternative techniques, anyone ever applied a notional rate of interest equal to the anticipated rate of profit.

For many of the simple techniques of the period it probably did not matter much if the entrepreneur neglected to impute a rate of interest on his locked-up capital. When the capital per unit of output was substantial, as it may have been in the construction of cotton mills, the need to impute some rate of interest would have been more evident, and probably rule-of-thumb methods were devised which corresponded well enough

[1] There is a summary of such data in Clark, op. cit, I, pp. 373–8.

in general effect with more rigorous accounting principles. It is possible that manufacturers reckoned the cost of their capital by the alternative uses to which they could put their savings. Over a country so large and diverse it is impossible to say much in general terms about the consequences of proceeding in this way. The opportunity-costs of industrial finance in America were clearly higher in the U.S.A. as a whole than in Britain. But the extremely high rates of interest which are sometimes quoted, for example on business paper, reflect a considerable degree of risk and other market imperfections, and therefore greatly exaggerate the disparity between the two countries.[1] While finance as well as labour was dear in America, the scarcity of finance attracted English funds more readily and earlier than the scarcity of labour drew out migrants. There were so few impediments to the import of British capital into America that the yields on long-term obligations in the two countries cannot have differed by very much more than the risk premium. An American textile manufacturer, asked by the Select Committee on Manufactures of 1833 to describe his methods of calculating capital-costs, said that he reckoned his interest upon the purchase price of the machinery and for this purpose took a rate of 6 per cent in America, and of 5 per cent in England.[2] The choice of these rates would have biased the American choice of technique towards the more capital-intensive methods.

In many cases, however, early manufacturers neglected to take account of interest except when they had to borrow from outsiders; each year they withdrew from the business enough to live on and ploughed back the rest irrespective of the yield on alternative methods of employing their funds. In effect, they behaved as if their capital cost them nothing. If entrepreneurs in both countries behaved in this way finance would certainly have appeared to be cheaper, in relation to labour, in America, than in England.

---

[1] For interest rates in the mid-West, see T. S. Berry, *Western Prices Before 1861* (Cambridge, Mass. 1943), pp. 411, 441, 513.

[2] P.P. 1833, VI, Q. 2617. When Gallatin wished to express as an annual figure the capital expended on barns in the Middle and Northern States he used a rate of 5 per cent. 'Suggestions on the Banks and Currency of the several United States', 1841, in *The Writings of Albert Gallatin*, ed. Henry Adams (Philadelphia, 1879), Vol. III, p. 254.

The nature of the spectrum of techniques

The practical importance of the inducements in the U.S.A. to adopt capital-intensive methods depended on the nature of the techniques available, and the possibilities they afforded for substitution between labour and capital. There clearly were several occasions on which one technique was manifestly superior for any likely range of factor-prices, and would therefore have been the most appropriate choice in England as well as America. The new techniques for spinning which were invented in the later eighteenth century were so much more productive for all factors than the old spindle that they were the best choice at any conceivable level of wages. But there were other situations in which the possible methods of production were sufficiently competitive, one with the other, for the manufacturers' choice to have been influenced by relative factor-prices. It is difficult to say how far the various new methods of spinning were substitutes for each other – Hargreave's jenny, Arkwright's water-frame, and Crompton's mule – and how far the suitability of each type of machine for the production of particular types of yarn specialized their uses; but, at some stages of development, some manufacturers may possibly have been influenced, in choosing between them, less by the market for particular grades of yarn than by the relative costs of the methods in the production of similar grades. In the years immediately after its invention the power-loom was not so decisively superior to the hand-loom that its adoption was uninfluenced by relative factor-prices; and even as late as 1819 it was not clear that in England the saving of labour was sufficient to outweigh the increase in capital-costs of the power-loom: ' . . . one person cannot attend upon more than two power-looms, and it is still problematical whether this saving of labour counter-balances the expense of power and machinery, and the disadvantage of being obliged to keep an establishment of power-looms constantly at work'.[1] On balance it seems reasonable

---

[1] Quoted by S. J. Chapman, *The Lancashire Cotton Industry* (Manchester, 1904), p. 31. Strassmann suggests that the improvements in the first half of the nineteenth century were predominantly labour-saving and capital-using. (W. P. Strassmann, *Risks and Technological Innovation* (Ithaca, 1959), pp. 118–19.) The improvements in flax-spinning machinery in the 1830s saved labour but required additional capital per unit of output (P.P. 1841, VII, Q. 3086–3092). A witness

to suppose that in the textile industry in the first half of the nineteenth century, the range of possible methods of production was sufficiently wide and continuous in respect of the proportions in which they used capital and labour for the choice of techniques to be responsive to relative factor-prices. And though the point could be settled only by detailed investigation, there is a general reason for expecting that similar conditions prevailed in other industries often enough to make labour-scarcity worth considering.

In the first place technical progress was still more empirical than scientific, that is it depended more on the response to particular and immediate problems of industrial practice than on the autonomous development of scientific knowledge. Technical development was therefore likely to take the form of slow modifications of detail, as opposed to spectacular leaps to a new technique decisively superior from the start to its predecessors; most even of the 'great inventions' of the period resolve themselves on close inspection into 'a perpetual accretion of little details probably having neither beginning, completion nor definable limits'.[1] (For the same reason the process of improvement was more likely to be sensitive to the factor-needs of the economy in which they were made.) In the second place a large sector of industry was organized on the domestic system. Under this system circulating capital was more important than fixed, and the commercial capitalist was always facing the question in what proportions to distribute his investment between fixed and circulating capital, that is, how much of his funds to lay out in raw materials to be worked up

---

before the Select Committee on Manufactures, Commerce, and Shipping of 1833 estimated that the efficiency of spindles had increased by 20 per cent since 1815, but that 15 per cent more machinery was required (presumably at the preparatory stage) 'to effect that improvement in quality which enables us to do that extra quantity'. This suggests that there had been no significant decrease in machine costs per unit of output of yarn over these years. (P.P. 1833, VI, Q. 5263.) The self-actor mule, in a case quoted in 1842, had higher capital-costs per unit of output than the hand-mule. (P.P. 1842, XXII, Factory Inspectors' Reports, p. 364.) Whitney's method of small-arms production required heavy and expensive machinery which was said to be worth installing only if it could operate for at least twenty years. (J. Mirsky and A. Nevins, *The World of Eli Whitney* (New York, 1952), p. 245.) This suggests that the method was capital-intensive in our sense.

[1] S. C. Gilfillan, *The Sociology of Invention* (Chicago, 1935), p. 5.

by domestic workers and how much on machines of his own. This choice was very sensitive to the cost of labour; and so long at least as the costs rose and indicated a shift *into* fixed capital, the commercial capitalist was in a better position to respond to the stimuli than his successors with a high proportion of their funds locked up in fixed capital. It was after all for an industry still organized principally on the domestic system that the Ricardo effect was postulated.

Moreover, even when the range of basic techniques likely to interest a manufacturer was very narrow or when one process was distinctly superior over a very wide range of relative factor-prices, it was possible to use them in a more capital-saving or a more labour-saving way, for example by varying the number of machines per worker, by running the machines for shorter or longer hours (by arranging workers in shifts) or at more or less rapid rates,[1] and by variations in the amount of space per worker or per machine.

The existence of methods of varying the factor-intensity of the basic techniques meant that there was usually a fairly continuous range of methods and that the method which used a little more capital saved a little more labour. But relative factor-prices would still be influential even if there were discontinuities such that, at some point, the alternative technique saved a great deal of labour but required a great deal more capital. Indeed the most striking disparities between English and American technology were probably established in just such cases. The gap between the hand-loom and the power-loom, and again, towards the end of the nineteenth century between the ordinary power-loom and the automatic loom, was

[1] A witness before the Committee on the Export of Machinery (1841), who had had experience of working in American cotton mills in the 1830s, said that cotton machinery was worked slower in the U.S.A. than in Britain, because the English employed more spindles per frame. (Q. 1830–1835.) This was also the opinion of George Wallis in 1854: 'The general speed of power-looms, and indeed of machinery generally, is lower than in England. By this means labour is economised, and one labourer can attend to more machines.' (P.P. 1854, XXXVI, p. 21.) Since running machinery more slowly is a method of saving labour at the expense of an increase in capital per unit of output, it is what we should expect in a country of dear labour. On the other hand Montgomery, who also had a detailed knowledge of the cotton-textile industry in both countries stated categorically that the Americans ran their spinning machinery faster than the English. For a possible reconciliation of this apparent conflict of evidence see below, p. 59, footnote 2.

wide. The capital-intensity of the 'superior' machine could be modified by running it longer, which some Americans did, but the need to make this modification is probably itself evidence that the new technique was much more capital-intensive than the old. The conditions of their labour-supply gave the Americans a much stronger incentive than the English had, to leap such a gap in the spectrum of techniques, with effects on subsequent technical progress which will be discussed later.

We have so far considered only the consequence for the choice of technique of the fact that labour was scarcer in America than in England, and have referred to natural resources only in so far as the abundance of agricultural land was a condition of the scarcity of labour. We must now take natural resources more explicitly into account.

The price of natural resources had an effect on the choice of techniques ranked by reference to the proportions in which they employed capital and labour. If natural resources were em-employed in the same proportions in the capital- as in the labour-intensive techniques, the price of natural resources would not affect the tendency of a rise in wages to shift the manufacturers' choice towards greater capital-intensity. If the supply of natural resources were inelastic, so that attempts to widen capital met rising costs for natural resources as well as for labour, the bias towards capital-intensity would be *strengthened* if the capital-intensive technique saved natural resources as well as labour, and *weakened* if it was more expensive in natural resources.

But this does not exhaust the possible effects of natural resources. For there may have been some alternative techniques, the principal difference between which was in their possibilities of substituting between natural resources and either capital or labour or both. Some techniques were important principally because of the proportions in which they used capital and natural resources: large blast-furnaces may have allowed a substitution of capital for raw materials, and the application of steam to water-transport may have involved the reverse sort of substitution – of power for capital; it is also possible that methods of building factories differed in respect of the proportions in which they used land and capital.[1] There were also techniques in which there was substitution between natural resources and labour; in particular there were possibilities of using power from water and steam instead of man-power.

Because of the unhomogeneity of natural resources and variations in their price within regions it is impossible to make any general statement about their cost, in relation to labour and capital, in the two countries. Land was certainly more abundant in the U.S.A. in relation to both other factors, and this fact dictated the choice of American agricultural techniques, which substituted land for labour. The abundance of land, and the nature of the American climate also enabled some substitution of natural resources for capital. There was less need than in England for investment in farm-buildings – the maize-stalks were left standing in the fields and they provided winter shelter for the cattle who were sometimes not brought in at all; and because of natural pasturage there was less need for winter feed. In some regions the type of agriculture was influenced by the ability to substitute natural resources for capital and/or labour: maize growing, for example, was a labour- and capital-saving, land-intensive form of agriculture. American agricultural methods which 'mined' the soil in effect substituted natural resources for labour and capital and so did the use of wooden frame houses. In industry too the lower rents for sites enabled New Englanders to economize in labour and capital in the construction of cotton-textile mills and also to build mills which enabled more effective use to be made of the textile workers and textile machines by allowing them more space. Similarly the American railways were built in ways which, in effect, substituted land for capital, as contrasted with the English railways which were built with a disregard for natural obstacles, a disregard which increased their engineering cost.

[1] The effect of lower rents for American industrial sites was to allow manufacturers to give greater weight to considerations which were favourable to ample factory space in all countries. 'One distinguishing feature of manufacturing establishments in the United States, both public and private, is the ample provision of workshop room, in proportion to the work therein carried on, arising in some measure from the foresight and speculative character of the proprietors, who are anxious thus to secure the capabilities of future extension, and in a greater measure with a view to securing order and systematic arrangement in the manufacture.' (P.P. 1854–5, L, p. 630).

Almost certainly also, power was cheaper, in relation to capital and labour, along the Fall Line and its supply more elastic than that available to some areas in England, and it may be that the mechanization of the Massachusetts cotton-textile industry was a substitution not so much of capital for labour as of cheap water-power for labour. If, in order to use cheap power, it was necessary to use more capital per unit of output, the high cost of labour gave an additional inducement to the substitution, but if the power was very cheap the substitution might have been profitable even had American wages been at the English level; and in support of the argument it might be pointed out that mechanization was much slower in sectors of the American cotton-textile industry, for example in Rhode Island, where labour was no less dear but where power was expensive. Moreover, water-power was, in effect, substituted for capital as well as labour.[1] The Americans ran certain types of textile machinery faster than the English, and this practice represented, to some extent, a substitution of natural resources for capital. 'Driving machinery at high speed,' wrote Montgomery, 'does not always meet with the most favourable regard of practical men in Great Britain; because in that country where power costs so much, whatever tends to exhaust that power is a matter of some consideration; but in this country (that is, the U.S.A.), where water-power is so extensively employed, it is of much less consequence.'[2] The American cotton-textile manufacturers also obtained their raw cotton on somewhat better terms than did the English, and this enabled them to economize in labour by using a better grade of cotton; in Lancashire manufacturers economized cotton at the expense of wages, using a great deal of short-stapled cotton.[3] In the construction of ships, cheap timber enabled American shipbuilders to economize in labour and capital.

This general line of argument is tantamount to the familiar

---

[1] As was wood where it could be used to provide power. The Americans developed high-pressure locomotives partly to overcome steep ascents and partly 'on account of fuel not being so much an object'. (P.P. 1841, VII, Q. 3000.)

[2] Montgomery, op. cit., p. 71.

[3] The greater need to make economical use of the raw material was one reason for the English preference for the mule-spindle, since the English used shorter-stapled cotton than the Americans and reworked more of the waste (Copeland, op. cit., p. 72).

view that the high productivity of American industrial labour was due principally to the fact that it was combined with richer natural resources rather than with more capital, though sometimes more capital per head may have been technically necessary to combine the labour with the resources.

But whatever force this argument may have for the later nineteenth century, in the period we are discussing it is not evident that, with the possible exception of cotton and wood, the natural resources relevant to industrial manufacturing were cheaper in relation to capital and labour in America than in Britain. Outside the Fall Line the supplies of power in the U.S.A. may well have been dearer and less elastic, in relation to labour and capital, than in England, since the supplies of coal were small; and until the discovery of new sources in the 1860s and '70s the same may have been true of iron-ore. In the manufacturing districts of New England, wrote Cobden 'the factory system has been planted under great disadvantages from the dearness of coal and iron'. Moreover the technical possibilities within industry of substitution between natural resources on the one hand and capital and labour on the other were less than the possibilities of substituting between capital and labour. We feel justified therefore in proceeding on the assumption that the dearness of American labour is the most fruitful point on which to concentrate in an examination of the economic influences on American technology.

LABOUR-SCARCITY AND THE RATE OF INVESTMENT

In the immediately preceding subsection we have explained how the dearness and inelasticity of supply of American industrial labour gave American manufacturers an inducement to adopt methods which were labour-productive even though they were capital-intensive. But these characteristics which explain the *composition* of investment would impose a restraint on the *rate* of investment. The shift to the more capital-intensive techniques partly offset the effect of rising labour-costs on the rate of profit, to an extent which depended on the adequacy of the existing range of techniques for the purpose. But the offset – in principle at least – could not be complete since, if the more capital-intensive method was as superior in productivity as this, it would already have been adopted in the U.S.A. before the rise in wages and in England at the lower wage-level. Once labour-costs had risen, the rate of profit would be higher with the more capital-intensive method than with the less, but it would nevertheless be lower than it was before the rise in the cost of labour, and, on reasonable assumptions, lower than in England; that is, other things being equal, the inducement to expand capacity would now be less than formerly and less than in England. Moreover, unless we assume (as was obviously not the case) that wage-earners were prepared to save the full increase in their earnings, the amount of new capacity that could be financed was reduced. So long as the volume of investment depended on the profit-rate, investment in the U.S.A. must have grown more slowly than previously for, since the machines in the more mechanized methods were more expensive, in terms of output, than the simpler machines, it must now have taken longer for the American manufacturers to accumulate the capital to produce a given output; and on these assumptions investment would also have grown more slowly than in England. In so far as the more capital-intensive equipment employed in the U.S.A. was produced in the U.S.A. the move to the capital-intensive end of the spectrum lowered the rate at which capacity was expanded to an even greater degree, because the capital-goods industry had to devote itself to producing equipment which was expensive in terms of output. To this extent labour-scarcity was a disadvantage which might partially be offset by substituting machinery, so far as this was technically possible within the range of existing methods, but which was none the less a disadvantage.

To put the matter in slightly different terms, the desire to widen capital was more likely in the U.S.A. than in England to run up against rising labour-costs. Hence the desire to widen capital was more likely to lead to its deepening in the U.S.A. than in England, deepening being the method of partially offsetting the fall in marginal profit-rates. But since, within the range of known techniques, the fall could not be completely offset, it imposed upon the desire of American manufacturers to widen capital a restraint to which English manufacturers were not subject.

This is the sort of situation envisaged by Marx. According

to him, labour-scarcity – the exhaustion of the reserve army of labour – would lead the capitalist to substitute machinery for labour, that is constant for variable capital; this would lead to a decline in the rate of profit, a fall in accumulation and in the demand for labour and a consequent replenishing of the supply of labour. The size of the reserve army was maintained by variations in the total of accumulation and in the proportion between fixed and circulating capital. With a smaller reserve army of labour, the tendency to substitute machinery for labour was stronger in America; but by the same token, the rate of investment was subject to a more severe restraint.

In some industrial activities the dearness and inelasticity of labour did in fact exercise a powerful restraint on the rate of investment. This happened even in an industry like the manufacture of small arms where techniques were available which, at first sight, might be supposed adequate to compensate for the high cost of labour. 'High wages,' wrote the Chief of Ordnance in a report to the Secretary of War in 1817, 'makes the business unprofitable to the contractors, and ultimately in many cases has occasioned their ruin.'[1] The true rate of profit on the manufacture of arms, when the capital-costs were accurately accounted, was low; and many concerns remained in business only because their primitive accounting concealed the fact that they were, in effect, treating capital as income and failing to provide for depreciation. It has been said of Simeon North, one of the inventors of the method of interchangeable parts, that: 'through the withdrawal as profits of sums which should have gone to pay for renewal charges, he squeezed his factory dry of its productive capacity and was forced after some years to start over again with new investment'.[2] For these, among other reasons, a very high proportion of the early small-arms manufacturers went into liquidation.

Nevertheless in a number of industries investment was rapid and in the industrial sector as a whole it is at least not evident that investment was slower in America in the first half of the nineteenth century than in England, despite the restraints of dear labour. It may be that the assumption which creates a problem out of this – the dependence of investment on the rate of profit on capital – is not valid; but for the moment we shall

[1] Deyrup, op. cit., p. 48. [2] Deyrup, op. cit., p. 54.

retain it, and try to consider what circumstances in America might have exerted a favourable influence on the rate of profit.

One possibility is that natural resources were cheaper and their supply more elastic in the U.S.A. than in England to an extent which offset the effects of its dearer labour. In this case the restraint which labour imposed upon accumulation in America would have been matched by a natural-resource restraint in England. But, as has already been argued, it is not evident that the natural resources most relevant to manufacturing industries were in fact cheaper in the U.S.A.;[1] and they would have had to have been considerably cheaper to offset the dear labour, since in most industrial products labour was a higher proportion of total costs than natural resources. We must therefore inquire in what ways dearness and scarcity of American labour might have favourably influenced the *rate* of investment. We shall consider three main ways.

In the first place, the American manufacturers had a greater inducement to organize their labour efficiently. The dearness of American labour gave manufacturers an inducement to increase its marginal productivity in all possible ways, and not merely in ways which involved the adoption of more capital-intensive techniques. The shortage of labour in America from colonial times encouraged prudence and economy in its use – Washington, for example, calculated with care the proper output of various types of labour on his plantation.[2] Americans from early times were often faced with a situation where a job had to be done – a house built or a river bridged – with the labour available on the spot, because the place was isolated and it was impossible to attract more labour. This gave them an enormous incentive to use their labour to most advantage, to make use of mechanical aids where this was possible, but in any case to organize the labour most effectively. Possibly lack of domestic servants led to an early rationalization of domestic duties and a corresponding increase in family efficiency; certainly the shortage of labour led generally to longer hours of work, to a general emphasis on the saving of time and a sense

[1] See pp. 44–7 above.
[2] R. B. Morris, *Government and Labour in Early America* (Columbia, 1946), pp. 38–9.

of urgency about getting the job done. In his account of his visit to America in 1818 Cobbett observed that 'the expense of labour . . . is not nearly so great as in England in proportion to the amount of the produce of a farm'.[1] The greater productivity of America, compared with British agricultural labour was partly the result of the fertility of the land; since labour was scarce, land which yielded a low return per unit of labour was just left uncultivated. It may also have been due to the avoidance of the more labour-intensive crops (for example dairy produce) as well as to the avoidance of labour-intensive soils. The superior physique and education of the Americans may partly have been responsible. Probably also, even in 1818, the American cultivator not only co-operated with superior natural resources, but had superior equipment. But Cobbett seems to suggest that the high productivity of American agricultural labour was in some measure due to the fact that its operations were more efficiently organized. The use of labour in English agriculture was much more wasteful than in English industry, partly from inertia and habit, partly because farm labour was so easy to get.

This labour- and time-saving pattern of behaviour was established on the farm from the early days of settlement – it was an ingrained attitude and not simply an economic calculation – and it was carried over into other activities. 'In England,' observed an English visitor to America in 1851, 'we cover our (railway) lines over with superintendents, police, guards, porters and a host of other officials; and relieve the passenger of many of those troubles which, in America, he contends with himself.' 'The American omnibus,' wrote the same author, 'cannot afford the surplus labour of a conductor. The driver has entire charge of the machine; he drives; opens and shuts, or "fixes" the door; takes the money; exhorts the passengers to be "smart", all by himself – yet he never quits his box.'[2] This attitude to labour was also carried over into industry and led to the more efficient organization of operations. H. C. Carey argued that the use of female labour in the American cotton-textile industry represented a more efficient use of the labour-force than was to be found in England, where men were used

[1] W. Cobbett, *A Year's Residence in the United States of America* (1818), p. 320.
[2] E. W. Wakin, *A Trip to the United States* (1852), pp. 130, 139.

for jobs that physically could be done by women if those women were given the right sort of equipment. In the U.S.A. the proportion of employed females to males was higher than in England – 'women being employed *here* (that is in the U.S.A.) because everything is done to render labour productive, while *there* (in England) a large portion of the power of the male operatives is wasted'.[1] The most conspicuous example of efficient use of labour is the training that the American manufacturers gave their workers so that each was able to handle more looms.[2] Whereas, in England, the weaver spent some of his time doing unskilled ancillary jobs, the American weaver did nothing but weave. The American arrangement probably involved a somewhat lower output per loom, that is, an increase in capital per unit of output, but there is little doubt that the English manufacturer would have found it profitable to adopt the same method of economizing labour. The point was that his need to do so was less; abundant labour, like the salt on the edge of the plate, tends to be wasted. 'Such a state of society where, as with us,' wrote the author of an English textbook on weaving in 1846, 'labour generally exceeds the demand for it, has a tendency to beget indifference to its improvement.'[3] In the manufacture of small arms, also, specialization of labour was carried much further in England than America, even before there were significant differences in technical processes; in England a workman specialized on one part of the weapons, but carried out all the operations on that

---

[1] H. C. Carey, *Essay on the Rate of Wages* (1835), p. 72.

[2] P.P. 1833, VI, Q. 5000. See T. M. Young, *The American Cotton Industry* (1901), p. 130. 'It is often found that a weaver will attend to four looms in the United States, who, in the same quality of work, would attend to only two in England.' [P.P. 1854, XXXVI, p. 21.] In 1860 the average was 4 per weaver in the U.S.A. as against 2 in Britain. In the 1880s the number of looms per weaver was 2 or, rarely, 3 in Germany, 4 in Lancashire, 6 (and sometimes 8 though at lower speeds) in Massachusetts. (Schulze-Gaevernitz, *Social Peace* (1893), p. 66 note. Copeland, op. cit., p. 10.) These differences partly reflect the nature of the product: more looms could be watched in the U.S.A. because there was a large demand for simple cloth, and the demand for the more elaborate, labour-intensive textiles was met by imports. Thus the differences in looms per worker was in part simply an international division of labour; but it also represents, and probably to a greater degree, the adoption in America of more capital-intensive methods of making similar products.

[3] George White, *A Practical Treatise on Weaving by Hand and Power Looms* (Glasgow, 1846), p. 331.

part – in the U.S.A. several workmen each performed only one or two operations on the part.[1]

Even, therefore, where there were no differences in technology or at least only such as involved the different disposition of identical machines, differences in the organization of operations may have ensured that the intensity or effectiveness of an hours' labour was greater in the U.S.A. than in England, and this tended to make up for the fact that the price of an hour's labour was higher in the U.S.A.[2] It is not an accident that scientific systems of labour management originated in America. Not only did mechanization and standardization make it easier to assess the effort needed for a particular operation, but the scarcity of labour made such assessment more necessary. In its turn careful management of labour bred careful habits in the worker; where labour was abundant it was wastefully used, and where it was wastefully used it was difficult for the worker to acquire the industrial virtues.

Secondly, dear labour not only provided an incentive to organize it more efficiently. It compelled American manufacturers to make a more careful and systematic investigation of the possibilities of the more capital-intensive of existing techniques.[3] Thus labour-scarcity could have had a favourable effect on the rate of investment by inducing the Americans to adopt, earlier and more extensively than the British, mechanical methods which would have been the most profitable choice even at the lower wages prevailing in England.

Labour-scarcity might, in the third place, have stimulated technical progress. Technical progress, that is movements of the technical spectrum as opposed to movements along it, would, by increasing manufacturing productivity, raise or at least keep up profit-rates, whether the progress was manna from heaven or induced by rising labour-costs. But manna from heaven one would expect to drop more readily in England, since England initially had much larger supplies of technical knowledge. The

[1] Deyrup, op. cit., p. 91.

[2] In civil employment 'we perform the same labour with a much less number of persons, whether officers or clerks, than in France'. (Gallatin, op. cit., p. 472.)

[3] The Committee on Machinery commented on 'the dissatisfaction frequently expressed in America with regard to present attainment in the manufacture and application of labour-saving machinery, and the avidity with which any new idea is laid hold of, and improved upon'. (P.P. 1854–5, L, pp. 630–1.)

point about labour-scarcity is that it constitutes a favourable influence on technical progress which was exerted more strongly in the U.S.A. than in England. Any manufacturer had an inducement to adopt new methods which made a substantial reduction of cost for all factors. But in their early stages, many of the methods devised in the nineteenth century could not be confidently assumed to effect such a reduction: before they had been tried out in practice for some time, estimates of their costs were highly conjectural. Where the best guess that could be made of a new method was that it promised a reduction of labour but some increase of capital, the Americans had a sharper incentive than the English to explore its possibilities. This is to say that labour-scarcity encouraged not only a careful and systematic investigation of the costs of the more capital-intensive of existing techniques, but the early adoption of any additions at the capital-intensive end which resulted from inventions of purely autonomous origin, even when they were made outside the U.S.A. Friedrich List wrote in the 1820s after a stay in America: 'Everything new is quickly introduced here, and all the latest inventions. There is no clinging to old ways, the moment an American hears the word "invention" he pricks up his ears.'[1] Montgomery, writing in the next decade about the cotton-textile industry, considered that though the number of specific inventions originating in the U.S.A. was not high compared with those that came from Britain, the common stock of inventions was very rapidly integrated into the American economy.

Labour-scarcity also gave Americans an incentive, not only to explore the labour-saving possibilities of autonomous inventions, but to attempt to invent new methods specifically to save labour. And if, as we shall argue later, the technical possibilities were richest at the capital-intensive end of the spectrum, the American was likely also to be better placed to make advances wherever, for any operation, he employed more capital-intensive methods than the English; that is the composition of American investment might have had a favourable effect upon its rate.

In the early decades of the century the principal effect of labour-scarcity in America was probably to induce American

---

[1] M. E. Hirst, *Life of Friedrich List* (1909), p. 35.

manufacturers to adopt labour-saving methods invented in other countries earlier and more extensively than they were adopted in their country of origin. The number of autonomous inventions was greater in the older industrial countries. But where their principal advantage was that they were labour-saving, they were more quickly adopted in the U.S.A. and labour-scarcity then induced further improvements, each additional improvement being perhaps small in relation to the original invention. And already in the early nineteenth century there were a number of important American inventions induced directly by the search for labour-saving methods and these became increasingly common as time went on.

Moreover, it was probably also easier for the Americans to adopt such methods. In England, where labour was abundant, labour-saving was likely to involve replacing, by a machine, labour that was already employed; in the U.S.A. it involved making a physically limited labour-force more effective by giving it machinery, but without displacing anyone, and with some increase in wages. There was, therefore, less opposition in America to the introduction of labour-saving practices and machines and of administrative methods for economizing labour: the fear of unemployment was less and the likelihood greater of gaining in higher wages from the increased productivity. In England, where there was a superabundant supply of hands and therefore 'a proportionate difficulty in obtaining remunerative employment, the working classes have less sympathy with the progress of invention'.[1]

For the same reasons, more changes in production methods came spontaneously from the workers in America than in England;[2] particularly when the worker had been self-

---

[1] P.P. 1854, XXXVI, p. 146. In the United States 'the workmen hail with satisfaction all mechanical improvements, the importance and value of which, as releasing them from the drudgery of unskilled labour, they are enabled by education to understand and appreciate'. For the attitude of English labour see *American and British Technology in the Nineteenth Century*, pp. 142–4.

[2] 'Every workman seems to be continually devising some new thing to assist him in his work, and there being a strong desire, both with masters and workmen all through the New England States, to be "posted up", in every new improvement, they seem to be much better acquainted with each other all through the trade than is the case in England.' (P.P. 1854–5, L, p. 38.) The Americans invented an excavating machine worked by steam, which cost £500 to £1,000, ran at £4 a day and did the work of 80 men. (P.P. 1841, VII, Q. 3024.)

C

employed earlier in life, and most of all when he had been a farmer, for he carried over into industry the inclination to seek his own methods of doing his job better. Thus in American canal-digging, the English methods were modified by the American farmers who devised a sort of primitive, horse-drawn bulldozer, similar to a device some of them had improvised on their farms. No improvement originated among the Irish navvies who dug the English canals.

If the methods adopted or developed by the Americans did no more than offset the initial disadvantage of high labour-costs, American entrepreneurs would have been on an equality with the English. In most cases, the methods must have done less than this. But in some cases they may well have done more. In exploring the borderland of blueprints, designs and embryonic ideas and hunches which lay beyond the end of the spectrum of existing techniques, it would not be surprising if the Americans hit upon some new methods which were so productive that they more than offset the high cost of labour, methods which reduced labour and capital per unit of output so greatly that they would have been the most profitable techniques even in the case of abundant labour. Very often the substantial reductions in cost came from ancillary developments and modifications made after the new technique had been operating for some time,[1] and these benefits accrued most fully to those who had adopted the method earliest; and the process tended to be cumulative, since the successful application of machinery to one field of activity stimulated its application to another, and the accumulation of knowledge and skill made it easier to solve technical problems and sense out the points where the potentialities of further technical progress were brightest.

Furthermore, quite apart from the effect of labour-scarcity on the incentive and ability to develop superior methods, the

---

[1] 'In fact it almost always happens that the inventions which ultimately come to be of great public value were scarcely worth anything in the crude state in which they emerged from secrecy; but by the subsequent application of skill, capital, and by well directed exertions of the labour of a number of inferior artisans and practicians, the crude inventions are with great time, exertion and expense, brought to bear to the benefit of the community'. John Farey, a prominent patent engineer, before the S.C. on the Laws relative to Patents for Invention. (P.P. 1829, III, p. 547.)

shift of American industry towards the more capital-intensive techniques provided the American machine-making industry with an active market which stimulated inventive ability among the manufacturers of machines and machine tools and perhaps also afforded it some advantages of scale. Ability to produce a labour-saving machine in one field also made it easier to develop machines in other fields. Thus the United States developed the typewriter, not simply because in America 'copying clerks could not be bought for a pittance' but also because in Remingtons, the Illinois gunmakers, there were manufacturers available who could put ideas into practical effect.[1] Standardization could be applied not only to final products but to the machines which produced them. 'Wood machines,' wrote an English expert, 'are made in America at this time like boots and shoes, or shovels and hatchets. You do not, as in most other countries, prepare a specification of what you need . . . but must take what is made for the general market.'[2] For these reasons there were cost-reducing improvements in the production of machines. Certainly by the middle decades of the nineteenth century there were some fields where the cost of the superior machines, relative to that of simpler machines, was lower in the U.S.A. than in Britain, and this was an independent stimulus to the adoption of more mechanized techniques in the U.S.A. There were also fields in which a superior machine was available for some operations in the U.S.A. but not in England.

Once a number of industries had been established in the U.S.A., a rise in real wages in any one of them due to technical progress exerted a similar effect on choice of methods as the initial high earnings in American agriculture. Where labour is scarce, any increase in productivity and real wages in one sector threatens to attract labour from other industries which have either to contract their operations or install new equipment which will raise their productivity sufficiently to enable them to retain their labour-supply.

In England where labour-supplies were abundant the technical progress in a single industry was not likely to stimulate

---

[1] J. H. Clapham, *An Economic History of Modern Britain* (Cambridge, 1938), III, p. 193.

[2] J. Richards, *Wood-working Factories and Machinery* (1873), p. 171.

technical progress in other industries by threatening their labour-supplies. It might, of course, stimulate technical progress in other industries by threatening their markets and in some cases their supplies of raw material; but not by threatening to draw off their labour. Any tendency for wage-earners within the technically progressive industry to establish a claim upon the fruits of their increased productivity was inhibited by the existence of a reserve army of labour, and the benefits of technical progress were likely to be diffused by means of lower prices over consumers as a whole, as in the case of the English cotton-textile industry.

In these ways the scarcity of labour gave Americans a keener incentive than the English had, to make inventions which saved labour. But it also gave them some reasons for being concerned with capital-saving. Throughout the previous argument the assumption has been made that the scarcity of labour biased the American entrepreneur's search for new methods towards those which specifically saved labour; since this is what contemporaries seem to assert.[1] But scarcity of labour, by exerting pressure on profits, did also provide some incentive to search for ways of economizing other factors as well. Contemporaries only rarely suggested that dear labour was a reason for saving capital, but Montgomery seems to have been arguing in this way when he wrote: 'the expense of labour being much greater in this country (the U.S.A.) than in Great Britain, the American manufacturers can only compete successfully with the British by producing a greater quantity of goods in a given time; hence any machine that admits of being

[1] 'A striking peculiarity in the drawing frames of this country (the U.S.A.) viz. their self-acting stop-motion, so far as I am aware, has not yet been introduced into the factories of Great Britain, nor do I believe it necessary that it should; because the helps in that country are very different from those in this. Here they are constantly changing, old hands going away, and new ones learning . . . In consequence of this continual changing, there are always great numbers of inexperienced hands in every factory; and as the drawing process requires the utmost care and attention to make correct work as well as to prevent waste, it is necessary to have the most expert and experienced hands attending the drawing frames; but this cannot always be obtained in this country as in Great Britain; hence it is more necessary to have some contrivance connected with the machinery here which will . . . prevent the work from being injured by inexperience on the part of attendants.' 'In Great Britain, where there is always a command of experienced hands, the introduction of this stop motion would be attended with no advantage.' (Montgomery, op. cit., pp. 57, 59.)

driven at a higher speed, even though it should exhaust the power, if it does not injure the work, will meet with a more favourable reception in this country than in Great Britain.'[1]

There is another link between labour-scarcity and attempts to save capital. When, from a given range of techniques, the American choice was more capital-intensive than the British, this in itself provided the American entrepreneur with an incentive to reduce capital-costs, in order to modify the large amount of machinery per operative; particularly where there were indivisibilities in the equipment, he needed to get more out of his machines in a given period of time in order to bridge gaps in the spectrum of techniques. He could do this, without any significant change in the technical characteristics of the machine, by running it longer and faster. Both these were methods of paying for the machine in a shorter period of time, that is of diminishing the interest bill on the cost of the machine and increasing the speed at which the amortization fund was built up. Because of the capital-intensity of their output, the Americans saved more on interest charges per unit of output than the English would have. (Thus, though running machines faster and longer is in effect substituting labour for machinery, it is usually a sign of a capital-intensive technique.) Montgomery observed in the 1830s that the Americans ran their cotton-textile factories longer hours than the English and drove their machinery at a higher speed 'from which they produce a much greater quantity of work'.[2] At the end of the century it was

---

[1] Montgomery, op. cit., p. 138.

[2] Montgomery, op. cit., p. 138. Two opinions have already been quoted to the effect that the Americans ran their machines more slowly than did the British (see p. 43 above). Montgomery himself said that no carding machines in America were driven at so high a speed as in England, and that generally they were driven at only half the speed. (Op. cit., pp. 32, 39.) In the passage quoted in the text above Montgomery seems to be referring to spinning machinery, though he does not explicitly limit his observation to this operation, and elsewhere (p. 162) he concluded that 'The factories of Lowell produce a greater quantity of yarn and cloth from each spindle and loom (in a given time) than is produced in any other factories without exception in the world'. It was certainly the general impression later in the century that the Americans ran their spinning machines faster. Mule for mule, said Young, New England produced more than Lancashire. One reason for the higher American speeds in spinning was the low marginal cost of power in the New England mills. But this was not the only reason. Spinning was a branch of the industry where American equipment was more capital-intensive than the English, and the Americans had therefore an incentive to run their

said of the American ironmaster that he 'wears out his furnaces much faster than the English ironmaster – in America furnaces require lining about every five years – and argues that the saving of interest on his fixed capital account justifies him in so doing'.[1]

But running machines faster and longer was only one of the ways of reducing capital-costs per unit of output. When the capital-intensive labour-saving machines had been installed, there usually proved to be possibilities of technical improvement in their construction and use. For the economy as a whole, one important form of capital-saving consisted of labour-saving improvements in the manufacture of machines, and to such improvements we can apply the previous argument about labour-saving improvements in general. The cost of machines in terms of output could also be reduced by improvements which increased their performance. Many of the inventions which were capital-saving in this sense were made as a result of attempts to improve machines whose principal advantage when they were first introduced was that they were labour-saving. In the textile industries there were few specifically capital-saving inventions. The initial effect of most of the great inventions was to save labour per unit of output at the expense of some increase in capital or at least without much saving. The saving of capital came later from such improvements as the increase in the number of spindles on each mule and the increase in the speed of the spindle. It was the manufacturers who installed the more complicated capital-intensive techniques who were in the most favourable position to make the subsequent improvements of this type.

---

machines at higher speeds, so long as by so doing they did not add disproportionately to labour-costs per unit of output. The effects of higher speed on labour-costs depended on the nature of the operations. When equipment was run faster and longer, more of certain types of labour were required per unit of output, for example the amount of piecing rose with the speed of looms, and the labour-cost of repairs increased; inputs of other types of labour remained the same, and of certain types fell. The proportions of these types varied according to industry. In processes where labour-costs rose rapidly with the speed of equipment, the Americans would not have much inducement to run their equipment rapidly. But it is conceivable that in some processes the incidental, net effect of speeding-up was to reduce labour- as well as capital-costs.

[1] S. J. Chapman, *Work and Wages; Part I, Foreign Competition* (1904), p. 87. The Americans also used larger furnaces.

There were other reasons why the Americans should have been anxious to save capital and in a favourable position to do so. In textiles, for example, American machines could be run faster because the marginal costs of water-power on the Fall Line rose less sharply than those of steam in Lancashire. Possibly too Americans wanted to get their money back sooner because they were readier than the English to assume that better methods would soon be available. These reasons will be considered later.[1] At this stage the point is only that labour-scarcity could lead to capital- as well as labour-saving; Britain on balance may have had stronger reasons for wanting to save capital, but America had some reasons which were not present to the same degree in Britain.

American manufacturers were readier than the English to scrap existing equipment and replace it by new, and they therefore had more opportunities of taking advantage of technical progress and acquiring know-how. This is a convenient place to consider the relationship of labour-scarcity to this American habit. In its extreme form the readiness to scrap is represented by Henry Ford who is reputed to have said that he scrapped existing machines whenever a new one was invented. But to judge from scattered instances and contemporary comment, this readiness was a characteristic of American industry very much earlier than Ford. The Secretary of the Treasury reported in 1832 that the garrets and outhouses of most textile mills were crowded with discarded machinery. One Rhode Island mill built in 1813 had by 1827 scrapped and replaced every original machine.[2] It would be difficult to parallel this from Lancashire. The English inclination was to repair rather than to scrap, and to introduce improvements gradually by modifications to the existing machines. John Marshall the Leeds flax manufacturer said in 1833 that his concern had 'reconstructed' its machinery twice in the forty-five years or so that he had been in business.[3] In the English textile industry as a whole it is doubtful whether equipment was often scrapped

---

[1] This point is discussed more fully on pp. 71–6.

[2] Strassmann, op. cit., p. 87. For evidence of rapid replacement see Clark, op. cit., I, p. 370, and Caroline Ware, *The Early New England Cotton Manufacture* (New York, 1931), pp. 135–6. For scrapping in the iron and steel industry see Burn, op. cit., p. 187.

[3] P.P. 1833, VI, Q. 2455.

except when a firm went bankrupt. The American readiness to scrap was noticed in other industries. One of the first English handbooks on woodworking-machines observed that 'there are throughout American factories but few wood-machines that have been running for ten years, and if any such exist there is a good and sufficient reason for abandoning them. The life of most machines used in joinery is not on an average more than six years . . .'[1] Contemporaries seem agreed on the general pattern of American behaviour: the American got something going, obtained his profits as quickly as possible, improved upon his original plant, and then scrapped it for something better. The problem is to explain his behaviour.

Scrapping is justified on strict economic grounds when the total costs of a new technique are lower than the prime costs of the old. The effect of high wages on decisions to scrap depended on the type of equipment being used and its age. The more mechanized and the newer American equipment was in relation to the English, the smaller incentive the Americans had to scrap in favour of a given new technique. This however does not exhaust the effects of dear labour on the incentive to scrap.

In the first place for a variety of reasons which we have already mentioned, the Americans tended to run machines faster and longer hours than the English; they also built them in a more makeshift fashion. In these circumstances, though capital per unit of output would still be higher in the U.S.A. than in England (otherwise it would have paid the English to adopt the same methods), the capital would be physically used up in a shorter period of time, and the American manufacturer would be in a better position to buy a new machine embodying the latest technique.

In the second place, American manufacturers seem to have expected a higher rate of technical obsolescence than the English. An American friend of de Tocqueville told him in 1832: 'there is a feeling among us about everything which prevents us aiming at permanence; there reigns in America a popular and universal belief in the progress of the human spirit. We are always expecting an improvement to be found in everything'.[2] While a rapid rate of achieved technical

---

[1] J. Richards, *A Treatise . . . on Woodworking Machines*, (London, 1872), p. 34.
[2] de Tocqueville, op. cit., p. 111.

progress is favourable to scrapping, expectations about technical progress have a more complicated effect. If the entrepreneur expects technical progress to be rapid, especially if he expects it to be more rapid than it has been in the past, if, that is, he assumes that the latest available technique will have a very short economic lifetime, its high average costs may prevent its adoption. In these circumstances the entrepreneur will put off the decision to scrap until a major technical advance appears, unless he believes that, in order to acquire the experience to take advantage of such a major advance, he must keep up with all the intermediate stages.

But the expectation of rapid technical progress had other influences which were more favourable to scrapping. The entrepreneur who expected new equipment to become obsolete, not so soon as to deter him from installing it, but sufficiently soon for him to want to ensure against the possibility, would not pay for durability. This is another reason for the flimsiness of much of American equipment. The American friend of de Tocqueville who has already been quoted said 'I asked our steamboat-builders on the North Bank a few years ago, why they made their vessels so weak. They answered that perhaps they might even last too long, because the art of steam navigation was making daily progress. In fact, these boats which made 8 or 9 knots could not, a little time afterwards, compete with others whose construction allowed them to make 12 to 15 knots.' The less optimistic expectations about technical progress among the English is one reason for the durability and heaviness of English machinery; no doubt the professional pride of machine-makers was mainly responsible, but if their customers had calculated on a rapid rate of technical obsolescence they would surely have been able to modify the prejudices of the engineers. There is another point. Because the Americans made their arrangements on the assumption that better methods would soon be available, they were more concerned to get back their money on new equipment as soon as possible. At the end of the century a French delegation to the Chicago Exhibition reported that American manufacturers invariably seemed to amortize their capital with the settled intention of replacing their machines by new and improved patterns.[1]

[1] E. Levasseur, *The American Workman*, ed. Th. Marburg (London, 1900), p. 61.

And this was probably the reason why earlier in the century Americans had the reputation of wanting their profits quickly. Thus technical, economic, and physical obsolescence were more likely to coincide in the U.S.A. than in England at least in those branches of activity in which American expectations about the rate of technical progress were most closely realized.

In the third place, in so far as technical progress took the form of inventing machines which saved labour, but with some increase (or at least no demonstrable saving) in capital-costs it might pay the Americans to scrap when it did not pay the English. The same considerations which warranted the American manufacturer shifting towards the capital-intensive end of the spectrum of existing techniques when he was adding to his equipment, might also warrant his replacing existing equipment when new methods were invented at the capital-intensive end. Where labour was abundant, and widening of capital could proceed at a constant wage, there was no inducement to replace existing equipment unless the new equipment yielded a higher rate of profit on the value of the old and new machines together. Where labour was scarce the preoccupation of the industrialist was with retaining or expanding his labour-force. His primary interest was with methods which would increase the productivity of labour and this was a more urgent concern than the return on capital, at least in the short run and so long as the return was enough to service any external finance and provide a conventional minimum return to the manufacturer. Accounting methods in the early nineteenth century were primitive – it was easier to calculate the likely labour-saving of a new process than its capital-costs. Manufacturers had to make their choice of technique on very rough-and-ready calculations on extremely inadequate data. The bias imported into the calculations by the nature of the labour-supply could therefore be the decisive factor. The American manufacturer was averse to retaining old equipment when more labour-productive equipment was available because the old equipment made poor use of his scarce labour. So long as the saving of labour was vouched for, the capital-costs were less important, at least within a fairly wide range, and in the absence of clear ideas and relevant data about the proper components of capital-costs, manufacturers were probably disposed to underestimate rather than overestimate them. But where, as in England, labour was abundant, and there was no pressing *need* to scrap, the calculations *had* to show, in order to warrant scrapping, a *higher* rate of profit on both machines than on the old equipment, and the results of calculations in these terms was almost inevitably biased against scrapping. The crucial difference is where the onus of proof rested: in America the presumption was in favour of any equipment which raised labour productivity; in England the presumption was in favour of existing equipment—the onus of proof was on the new equipment, it had to be demonstrated that it would yield a higher rate of profit.

The fact that maintenance-costs were mainly labour-costs, and that they tended to increase rapidly with the age of equipment, reinforced the American inducement to scrap; the costs of keeping a given piece of equipment intact were greater than in England.[1]

Given the high costs of labour, and the inadequacy of existing accounting concepts and data, the readiness of Americans to give new labour-saving methods the benefit of any doubt about their capital-costs was a rational one. Even so, in particular cases, it may have led to scrapping in circumstances when it was not justified; the scrapping of old machines and the installation of new ones must sometimes have involved wastes which only capitalists who enjoyed superiority in respect of some other factor could afford to bear. It has been suggested for example, with particular reference to the replacement of horse-drawn trams by electric trams, that the greater readiness in America than in Europe to discard equipment may have been due to an inadequate analysis of the costs of change.[2]

But American readiness to scrap was partly unrational from an economic point of view. The expectation of more rapid technical progress was to some extent the result of the general optimism of the American character, and was initially independent of economic facts, rigorously defined. The Americans were, as Cobden said, a 'novelty-loving people'. In so far as

[1] An English machine manufacturer, in evidence before the R.C. on Export of Machinery (1841) said that repair required more skill than manufacture (Q. 3112).

[2] H. Jerome, *Mechanisation in Industry* (Nat. Bureau of Economic Research, 27; New York, 1934), p. 333.

the decision to scrap was taken from mere love of the latest method, it was even more likely to let down those who acted on it. Even in such cases, however, the decision, though 'unrational', may have turned out to be warranted by the eventual course of technical progress. Ford's readiness to scrap whenever a new machine was available might be reconciled with the orthodox criteria by assuming that, though it could not be reliably predicted of the new machines *before* their introduction that they would reduce costs to the requisite to justify scrapping, in more cases than not they proved to do so in practice. De Tocqueville's informant claimed that this was true of American ship-building, for he continued, 'And in fact this (that is, the expectation of improvement in everything) is often correct'.[1] Possibly also – a variant of the same explanation– the constant pursuit of the latest invention may have led to the adoption of machines which, though not themselves economic, ultimately put those who adopted them in a position to take advantage of later inventions which made spectacular reductions in cost. If, as is argued later, the more capital-intensive of the existing range of techniques had the greatest possibilities of technical progress, a persistently optimistic view of the costs of new labour-saving methods led in the long run to the accumulation of experience and to further technical progress which outweighed the waste of capital into which it sometimes led American entrepreneurs.

This technical progress was not without its costs. The pursuit of the latest method must sometimes have dissipated a firm's resources without yielding a commensurate increase in experience. Possibly there was too much imitation of the successful leaders, and capital-intensive methods were adopted by some concerns which would have been better advised to close down. Expectations about technical progress were not always realized, and equipment which could have been made more durable at modest expense when first installed had to be renewed at greater expense because it had fallen to bits well before new methods became available. Sometimes equipment was so makeshift that it quickly deteriorated, worked very inefficiently, and was costly in repairs. But it seems reasonably clear that on balance and over the economy as a whole, the

[1] de Tocqueville, op. cit., p. 111.

American habit of not building the equipment to last and the closely-associated readiness to scrap were favourable to growth. For they meant that the American capital stock tended to be younger than the English and to embody more technical knowledge. Once achieved technical progress in any line became more rapid in America than in England, this in itself weighed strongly in favour of earlier scrapping; the point we have now been making is that the American propensity to scrap developed *before* technical progress was more rapid in America, and is therefore to be included among the independent sources of such progress. In this field, if not in others, what entrepreneurs expected came to pass because they expected it.

The scarcity of labour may therefore have exerted favourable influences on the rate of investment by inducing (i) a more efficient organization of labour, (ii) a more rapid adoption of autonomous inventions, (iii) a higher rate of technical progress and (iv) greater readiness to scrap existing equipment in order to take advantage of technical progress. There is also another way in which the composition of American investment might have exerted a favourable effect on its rate. The propensity to save out of profits may have been a function of the degree of capital-intensity. A plausible case can be made out for this view. A man who runs a machine is more likely to be interested in the possibilities of mechanization than the man who runs a sweat shop. The industrial capitalist whose capital was tied up in his factory and its equipment was more concerned with technical development than the commercial capitalist under the domestic system – and indeed the gain, in the early days, from any shift from the domestic system to the factory may have arisen principally from the increasing role it assigned to the type of capitalist who was interested in technical possibilities. 'The threat of obsolescence and the attractiveness of new and better machines make the capitalists with expensive machines more accumulation-minded than entrepreneurs with little capital.'[1] If this was so in America in the first half of the nineteenth century, even though the resort to more capital-intensive methods did not prevent a fall in the rate of profit on

[1] Albert O. Hirschman and Gerald Sirkin, 'Investment Criteria and Capital-Intensity Once Again', *Quarterly Journal of Economics*, LXXII, August 1958, p. 470.

capital, the effect of this fall on accumulation may have been partially offset by a rise in the proportion of profits reinvested.

We can now summarize the argument to this stage. The dearness of American labour and the inelasticity of its supply provide an adequate explanation of why, from a given range of techniques, the choice of the American manufacturer should have been biased towards those which were more productive per unit of labour because they were more expensive in capital per unit of output. The same circumstances might also have exerted favourable influences on the rate of investment by providing an incentive to devise new labour-saving methods, and because capital-intensity of investment increased the ability to devise such methods and also increased the propensity to save out of profits. The main point is the favourable effect of labour-scarcity on technical progress. This might resolve the sort of dilemma emphasized by Marx.[1] It is often said that this dilemma was resolved by autonomous technical progress which sustained the rate of profit. The implication of the argument which we have been pursuing is that technological progress might itself have been the result of the exhaustion of the reserve army and not something introduced from outside the system.

The difficulty about this theory, as of any theory which regards 'restraints' as a net favourable influence on growth, is that it does not, in itself, explain why American manufacturers should have been able and prepared to continue investing in capital-intensive methods to the point where this investment yielded rapid and substantial gains in technical knowledge. Shortages frustrate as well as stimulate.

EXPANSION OF THE MARKET

In most of the preceding discussion we have assumed that the rate of industrial investment was limited by the supply of factors and depended on the rate of profit. We have assumed that, via the rate of profit, the level of investment in both the U.S.A. and in England was adjusted to the rate of increase of supply of all factors, and we have considered the consequence of the fact

[1] Marx himself argued in chapter 13 of *Capital* that manufacturers were induced by legislative restrictions on hours of work to look for mechanical improvements; but he did not regard the state of the labour-market as a strong impetus to technical progress. See p. 48 above.

that, when investment increased at these rates, shortages of one particular factor, labour, appeared earlier in the U.S.A. than in England. But the rate of investment also depended on the expansion of the market. We must now consider the possibility that, in relation to the total supply of factors to industry, the demand for American manufactures in the first half of the nineteenth century was expanding more rapidly than the English.[1]

This is prima facie probable. The major part of the demand for American manufactures came from the rise in agricultural incomes as cotton exports from the southern states rose and as the country was opened up and population increased.[2] In contrast, though a substantial part of the demand for English manufactures also came from an increase in consumers' incomes in primary producing-areas, a considerable part came from a switch of demand from domestic-type industry to English factory industry, and depended on a fall in the cost of the English products. Thus the long-term growth in the demand for American industrial goods probably warranted a rate of investment, in relation to the supply of factors as a whole, which was more rapid than that in England.

Moreover, it is probable that any given increase in demand was likely to lead to a larger increase in investment in America than in England. In England an increase in demand was met from the existing centres of production where there were generally some possibilities of increasing output with small changes in existing equipment. In the U.S.A., because of the imperfections in the product-market which we have already discussed, an increase in demand in a new area of settlement was likely to be met by the creation of new capacity and new concerns within the area, even though there may have existed

[1] This presumably is what contemporary observers like Gallatin (op. cit., p. 430) meant when they said that American development was retarded by both dear capital and dear labour.

[2] In the early stages of industrialization the expansion of exports of primary products from southern states was of critical importance. It is unlikely that these export staples could have been profitably produced by dear white labour. 'In a country where labour is as expensive as it is in America,' wrote de Tocqueville, 'it would be difficult to grow tobacco without slaves' (op. cit., p. 77). Thus in a sense the industry of the northern states was able to escape the consequences of dear labour by mechanization because the southern states, where crops were labour-intensive and mechanization would have been technically difficult, had been able to escape the consequences of dear labour by importing it.

some slack in the older area, and without corresponding dis-investment in the older areas.

The rate of investment in relation to total factor-supplies is relevant to our argument in two ways. It is relevant, first, to the effect of factor-supplies on the invention and adoption of new methods. To the extent that investment was pressing more closely on total factor-supply in America than in England, the advance of technology would have been more sensitive to factor-endowment in America. Technology may be expected to edge along, adjusting itself to relative resource-scarcities, only when there is, to begin with, a rough balance between resources and investment – when the increase in capacity is constantly pressing on the available supplies of labour, natural resources and finance. It is in these circumstances that manu-facturers are most likely to get clear indications of relative resource-scarcities, that those who adopt methods which are inappropriate to the relative resource-scarcities of the economy are penalized and that a new technique – whatever resource it saves – is likely, when introduced into one firm or industry, to have repercussions on other firms or industries and force them either to contract or innovate. Thus the firms in the U.S.A. which in the 1820s and '30s adopted labour-intensive tech-niques would have been placed in a disadvantageous position in competition with firms in the same or competing industries which had made a more appropriate choice and had adopted capital-intensive techniques. But this was not true to the same extent of firms which adopted capital-intensive techniques in England, the main effect of which would simply have been to depress wages. For the same reason new techniques which increased the productivity of all inputs had a more general effect in the U.S.A. than in England. If, in the early decades of the century, the actual rate of growth was higher in the U.S.A. compared with the possible rate of growth, this fact must be counted as a circumstance favourable to technical improve-ment, quite apart from the relative scarcity of labour *vis-à-vis* other factors.[1]

---

[1] Though heavy demand might sometimes blunt the edge of the incentive to improve technique, 'The cotton-textile machinery industry of the 1870s had been softened by three or more decades of heavy demand . . . in which the advance in machine technology had been very slow'. (Navin, op. cit., p. 109.)

The fact that investment in America was rapid in relation to the supply of all factors except agricultural land gave an incentive to improvements which were capital-saving as well as to those which saved labour. Where the existing techniques afforded no possibilities of alleviating the dearness of labour but the entrepreneur still persisted in the attempt to widen, the difficulty of getting capital would induce him to concentrate on making the most of his machinery in all ways which did not involve the use of additional labour per unit of output. Where the range of existing methods *did* allow the entrepreneur to compensate for dear labour by substituting capital, the adoption of more capital-intensive techniques would provoke the problem of finance even more acutely, especially if the techniques were very effective in saving labour. When investment is pressing against resources as a whole, the temporary resolution of the most severe scarcity is likely to be followed by the emergence of scarcity in some other factor.

A 'scarcity' of capital not only provided the user of machines with an inducement to get the most out of them: it gave the manufacturer of equipment an inducement to provide them as cheaply as possible. Many of the goods, the manufacture of which was most highly mechanized, were not single-use consumer goods but equipment designed to increase labour productivity or at least to meet a production problem (like the steel ploughs which were necessary to open up prairie soils). Where the cost of the minimum feasible piece of such equipment was large in relation to the funds available to the typical user, the demand was very sensitive to its price. This was said to be the case with woodworking-machinery[1] and it was probably also true of some types of agricultural machinery, of such goods as sewing-machines and of such activities as ship-building. This was one reason for the flimsiness of much American equipment. According to the American friend of de Tocqueville, whom we have already quoted, 'one reason why our ships do not last long is that our merchants often have little capital at their disposal to begin with. It is a matter of

---

[1] Richards, *Treatise*, op. cit., pp. 35, 50–51. 'The (woodworking) machines are for the most part sold to men of limited means, who have not only to consider the worth of the money after investment, but have first the greater difficulty of commanding a sum sufficient to purchase the machines.'

calculation on their part. Provided that the ship lasts long enough to bring them in a certain sum beyond their expenses, their aim is attained'.[1]

Almost every observer pointed to this contrast between the durable English and the flimsy short-lived American equipment. For example, writing of woodworking-machines, an English author observed 'For builder's machines to supply the American market they must be cheap, capable of doing the maximum amount of work when operated with first-class skill. . . . For the English home market machines must be better fitted to correspond to the general character of other engineering work. Changes not being so frequent, and the first cost less, they can be made heavier and stronger'.[2] For this general contrast there are many reasons. The flimsiness of American construction may have been partly the result of the technical inability of American engineers to make high-quality durable machines, and to the high cost of iron and the low cost of wood. It was partly the result of the fact that American engineers were not so long-established a profession, applied less rigorous technical standards and so allowed more weight to be given to economic considerations, in contrast to English engineers who were apt to subordinate economic considerations and who sometimes boasted of the fact. 'We would not be understood,' wrote the author we have just quoted, 'as making the plans and designs of the engineer subservient to commercial conditions.'[3] If British purchasers of machines had consulted exclusively economic interests, unconfined by the technical prejudices of engineers, they might have preferred cheaper and less durable machines. But in addition there were quite rational economic reasons why Americans should have attached more importance than the English to building cheap machines of the sort which would bring quick returns. There was the need discussed earlier to modify the burden of the large amount of capital involved in the choice of capital-intensive techniques; though the American chose the more capital-intensive of existing techniques, any given technique was apt to be embodied in a less durable form in America than in England. This need made the American purchaser of

[1] de Tocqueville, op. cit., p. 111.   [2] Richards, *Treatise*, p. 53.
[3] Ibid., p. 51.

machines readier to accept standard products, and less inclined to force upon machine-makers minor, but expensive, modifications. Then there was the expectation of more rapid technical obsolescence. But all these reasons had greater force in so far as investment in America was pressing closely on capital as well as on labour.

In building transport-systems, too, the Americans attempted to economize capital more resolutely than did the English. Edward Watkin, a leading English railway-builder, put the contrast between the methods of the two countries at the middle of the century in the following terms:

> The cost of American lines has been brought down by the necessity of making a little capital go a long way, and by the sacrifice of many of the elements of permanent endurance which attach to our railways. We have deemed the inventions of railways a final improvement in the means of locomotion, and we have, therefore, constructed our works to last 'for ever' of bricks and mortar. We have made our rails strong enough for any possible weight of engine; our drainage capacious enough to remove any conceivable flood . . . our bridges firm enough for many times the weight that can ever come upon them.

The Americans believed that the English desire for permanence was 'a bar to future improvement; while their [the American] plan for putting up with "what will do" leaves the door open for invention'.[1] These were typical of the views held by most people who had experience of the two countries. 'In making railways in the United States,' said an American witness before the Committee on the Export of Machinery, 'we aim to economize capital, and we therefore are not so particular about reducing the gradients as you are.'[2]

> In the construction of railways [said another observer], economy and speedy completion are the points which have been specially considered. It is the general opinion that it is better to extend the system of railways as far as possible at once, and be satisfied in the first instance with that quality of construction which present circumstances admit of,

---

[1] Watkin, op. cit., p. 124.   [2] P. 1841, VII, Q. 3012.

rather than to postpone the execution of work so immediately beneficial to the country . . . .

Whitworth quotes the case of the railway connecting the east and west parts of Pennsylvania, which in 1854 was carried over the Allegheny Mountains by a series of five inclined planes, operated by stationary engines. A new continuous road was then under construction which would dispense with the inclined planes and save four hours. 'It is doubtful whether the delay would not have been very considerable, had the construction of the railroad been postponed until means had been found for executing these great works in the first instance.'[1] The American canals seem also to have been built quickly and cheaply. 'They have built the longest canal in the world in the least time for the least money' wrote one observer about the Erie canal.[2]

At least in the case of railroads, however, there are some explanations of flimsy and rapid American construction which are independent of desire to economize on capital. Thus, over its entire lifetime, an American railway was likely to be less intensively used than an English railway and therefore did not need to be so well built. There was also in America a much greater disparity between intensity of usage in the immediate future and in the longer run. In America a system big enough to carry the load expected when the new region was fully opened up would be much too big for the traffic in the years immediately after building; and therefore the railways were built in ways which allowed them to be modified most easily when the increased demand came.[3] The English, who could

[1] P.P. 1854, XXXVI, p. 127.

[2] C. Goodrich, *Government Promotion of American Canals and Railroads, 1800–1890* (New York, 1960), p. 153.

[3] The prospect of future expansion also influenced methods of railway construction through its effect on the railways' financial policy. Except in Massachusetts, New York State and Pennsylvania, 'in any majority of cases the (financial) policy pursued is to extract a dividend at the earliest possible moment; to pay that dividend to the last farthing of available surplus over working expenses; and to trust to the increase of traffic for providing, when the emergency may arrive, for the deterioration of the permanent way, and the rebuilding or replacement of worn-out stock. Very few companies, indeed, systematically provide for the renewal of the perishable parts of their property by a reserve fund, regularly put aside out of the annual revenue; and the reason given for the omission is, simply, that by the time permanent way requires renewal, or timber bridges and viaducts

expect the load soon after building to be not dramatically below its maximum, had no reason not to build to last from the start. Rapid building was also prompted by the desire to obtain strategic advantage in relation to other competitors, which involved covering as much territory as possible despite limited resources; and by the wish to qualify for land-grants, which were a function of miles built. In these ways some of the differences between American and British railway-engineering can be explained without supposing that the Americans were more interested in saving capital per unit of output. But it is probable that an additional reason was that a higher rate of return on capital was expected of public utilities in the U.S.A. than in Britain, and this would result in lower capital-intensity. British canal- and railway-builders tended to project on the basis of 5 or 6 per cent rather than the industrial rate of profit. The Americans required more, and even 1 or 2 per cent more could make a substantial difference in projects in which capital is important in any case.[1]

In a number of instances, therefore, the rapid rate of American investment, in relation to factor-supplies as a whole, was a reason for American interest in methods of economizing capital, over and above the reasons previously discussed.

The long-term growth of the market as a result of rising incomes in agriculture and the filling-up of the country is relevant to the argument in a second way: it sustained the incentive to widen capital. According to the argument in an earlier section the level of real wages in American industry was set by the absolute level of investment in relation to labour-supplies, subject to the minimum below which real wages in manufacturing could not go, a minimum determined by earnings in agriculture. The desire to widen capital – because

---

replacement, the traffic on the lines will have become so greatly developed that it will be then more advisable to enlarge the capital to the extent required, and replace and renew everything on a more permanent and substantial scale.' (Watkin, op. cit., p. 133.)

[1] In the later 1860s the average rate of interest paid by railway companies in the United States amounted to 8 per cent per annum; in England the rate of interest on long obligations was not over 5 per cent. As a result of this difference it paid to substitute dear (but more durable) steel for iron rails at an earlier stage in England than in America. Abraham S. Hewitt, *Selected Writings*, ed. A. Nevins (New York, 1937), p. 31.

of the protection of tariffs and high transport-costs and the filling-up of the country – which determined the level of investment, came up against rising marginal labour-costs, and therefore led to capital-deepening, in order partially to offset the fall in marginal profit-rates. But, within these assumptions, the fall in marginal profit-rates could not be completely offset and would therefore reduce the desire to widen capital which was the impulse which set the whole process in motion. If one introduces technical progress, whether autonomous or induced by the fall in profit-rates, this would help to keep up profit-rates and might raise them, thus maintaining the desire to widen. The relevance of the long-term growth in demand is that it would help to sustain the level of accumulation for normal widening-purposes even in periods when technical progress was sluggish. The growth in demand, that is, contributes to resolving the problem of reconciling the rate of accumulation with the bias towards capital-intensity. This explanation would not be inconsistent with those offered at an earlier stage in the argument but it might render some of them otiose.

2 The Genesis of American Engineering Competition, 1850–1870[1]

D. L. BURN

[This article was first published in *Economic History*, Vol. II (January 1931).]

The view that England was industrially without a serious rival till quite late in the nineteenth century needs more qualification than it usually receives. There were directions in which the revolution of industrial method and organization was by the middle of the century proceeding more rapidly in the United States than in England.[2] Professional recognition of this by engineers was widespread though not universal from 1850; but it has been the subject of rediscovery by successive generations of journalists,[3] and its early phase, between 1850 and 1870, has been neglected by English historians. In these years it was illustrated by the very frequent adoption of American methods in English industry, and to a much smaller degree by effective competition in neutral markets. The character of the American advance was brought into prominence at the Industrial Exhibitions which were fashionable from 1851, and by good fortune it was analysed in the fifties by several distinguished observers, notably by Joseph Whitworth.

Some of the most interesting of these early English records of American progress come from the visits of two small Commissions to the industrial areas of the United States. Unexpected American advances were recognized at the Exhibition of 1851, and when it was decided to hold a similar

---

[1] My thanks are due to two members of the Engineering School at Cambridge, Mr T C. Wyatt, Fellow of Christ's College, and Mr G. E. C. Gordon, Trinity College, for assistance in some technical problems; and to Mr C. W. Guillebaud, Fellow of St John's College, for some valuable suggestions, of which I have made use in this article.

[2] This may to some extent be traced in modern surveys of American industrial growth, particularly in V. S. Clark, *History of Manufactures in the United States*.

[3] Cf. *American Engineering Competition*, reprinted from *The Times*, 1901.

exhibition at New York in 1853–4 the English Government sent distinguished commissioners to report. When they arrived they discovered that the Exhibition was not assembled at the advertised opening date, and they determined to substitute tours in the American industrial districts in place of the inspection of the new Crystal Palace. The commissioners drew up reports on their tours, and two of them, those of Joseph Whitworth and George Wallis, give a valuable sketch of New England manufactures.[1]

A similar tour was instituted shortly afterwards by the Ordnance Department, with a view to discovering how far American methods might be adopted to increase the output of small-arms. The moving spirit in this was John Anderson, Inspector of Machines in the department, who was influenced by the methods employed by Colt, the American pistol manufacturer,[2] and by Whitworth. Anderson had recently toured England to investigate productive methods, and he was a member of the Commission sent to the States. He and his colleagues 'deemed it advisable to allow as little as possible connected with the private manufactures of the country escape them, and visited many works which at first sight may seem totally irrelevant to the branches of manufacture to which their attention was more particularly directed.' They subsequently presented a report which covered much the same ground as those of Whitworth and Wallis.[3]

The three reports were drawn up by men obviously well qualified to judge the novelty and efficiency of the methods they discussed, though likely in the circumstances to describe the best practice rather than the average. In all cases the writers emphasize 'the eager resort to machinery wherever it can be

[1] *New York Industrial Exhibition; Special Reports of Mr George Wallis and Mr Joseph Whitworth*, P[arliamentary] P[apers], 1854, XXXVI. Abbreviated in the notes as *Wallis* and *Whitworth*. The pagination adopted is that of the volume of bound reports, not that of the reports as issued individually. George Wallis (1811–91) started life as a practising artist, but turned his attention to the application of art education to manufacture, holding several important positions from 1841. He was Superintendent of the textile divisions of the London Exhibitions in 1851 and 1862; appointed Headmaster of the Birmingham School of Design in 1852, and Senior Keeper of the Art Collections at South Kensington in 1858.

[2] *Vide infra.*

[3] *Rep[ort of the Commission] on [the] Machinery [of the United States*; P.P., 1854–5, L.]. Apart from Anderson the Commission included Lt.-Col. Robert Burn, R.A., and Capt. T. P. Warlow, R.A.

applied', and the extent to which industries were organized for mass production in large factories, where in England outwork and handicraft persisted. Though they often criticized the quality of machine tools where there were English parallels, they had nothing but praise for the adoption in most branches of industry of the 'manufacturing principle', the production in large numbers of standardized articles on a basis of repetition in factories characterized by ample workshop room and 'admirable system', designed to assist the progress of materials through the various stages of production. The principle lent itself to the use of mechanical methods, particularly to the development of automatic 'special purpose' machines designed for a single operation, and dispensing largely with skilled labour. In devising machinery of this type 'the Americans showed an amount of ingenuity, combined with undaunted energy, which we would do well to imitate if we meant to hold our present position in the great markets of the world'.[1]

The commissioners gave a wealth of illustration of the tendencies they noticed. All were impressed by the woodworking industries. Whitworth discovered 'many works in several towns occupied exclusively in making doors, window-frames, or staircases by means of self-acting machinery such as planing, tenoning, mortising and jointing machines', whereby builders were supplied with goods cheaper than they could produce in their own shops.[2] Such factories were not necessarily large – one with a daily output of 100 doors employed 20 men.[3] The same method was applied to the manufacture of furniture, everything 'from sawing to sandpapering' being often performed by machinery.[4] So too with agricultural implements – ploughs and mowing-machines. Whitworth drew attention to a factory 'where all the ploughs of a given size are made to the same model, and their parts . . . are made all alike'.[5] In the manufacture of the bigger metal products less specialization was found than in England; but frequently the American method of making smaller metal goods was the more advanced. Wood screws of a better pattern than those made in England

---

[1] *Rep. on Manufactures*, pp. 578 and 630–31. *Whitworth*, p. 41; *Wallis*, pp. 12, 13.
[2] *Whitworth*, p. 116.
[3] Ibid.
[4] Ibid., p. 118, and *Rep. on Machinery*, p. 613.
[5] *Whitworth*, p. 118.

were produced by automatic machines several of which could be supervised by one person.[1] Files were machine-made, with one supervisor to two machines; the reduction in price more than compensated for some loss of quality, and the product was satisfactory in the view of the English commissioners; though the best files were still imported from England.[2] Cut nails were machine-made in England at this time, but the commissioners were all impressed by the discovery in several towns of 'immense nail manufactories' – one described by Whitworth employed 250 workpeople – and the scope of the machine work was in some respects novel.[3]

In the manufacture of locks, clocks, and small-arms machine methods were applied in more exacting conditions, where the accurate fitting of components was essential. One factory at Newhaven – not an isolated example – with 200 employees was devoted entirely to the manufacture of locks and padlocks; 'special machines were applied to every part, and the parts of locks of similar description can be interchanged'. The output was 2,000 per day, some of good quality being sold at $2\frac{1}{2}d.$ each. Whitworth wrote that padlocks produced here were 'of a superior quality to those of the same class ordinarily imported from England, and not more expensive'.[4] Jerome's clock factory in the same town, representing one of the great American industrial achievements, was also described. With a staff of 250, many being boys and girls, it had a daily output of 600 clocks, and an established reputation in English and continental export markets; and it could produce a clock for a dollar.[5] The commissioners, however, gave most detailed attention to the manufacture of small-arms. It was found that in the production of muskets at the Springfield armoury 'the parts were so exactly alike that any single part will in its place fit any musket', even if the parts were made in different years.[6] In achieving this result there was extraordinarily little hand-work. The sixteen operations needed for the production of the wooden

---

[1] *Rep. on Machinery*, p. 556.

[2] Ibid., p. 617.

[3] Ibid., p. 615; *Whitworth*, p. 111.

[4] *Whitworth*, p. 115; *Rep. on Machinery*, pp. 558 and 616.

[5] Ibid., p. 616; *Whitworth*, p. 115. Whitworth shows considerable interest in the stamping presses employed.

[6] Ibid., p. 130.

stock involved two and a half minutes' hand-work and twenty-nine minutes' machine-work.[1] The metal-work was performed mainly by 'circular milling tools' devoted to a single operation, and among the machines were some in which a succession of tools were brought to bear on the work one after the other (with the 'most rigid accuracy'), and 'edging machines', to 'trace irregular figures and impart an exact outline', the form being taken from a model.[2] Such developments of the milling machine were quite foreign to English engineering. Springfield was also distinguished for the saving of labour in moving material about the factory, lifts and carriages being employed.[3]

The commissioners were not as pleased with the American mode of making machines as with their resourcefulness in using them. Whitworth found that the 'engine tools' were 'similar to those in use in England some years ago, being much lighter in construction than those now in use, and turning out less work in consequence'; though significantly enough he discovered a greater proportion of lathes with slide rests.[4] Similarly, the Ordnance officials discovered that in the States the types of machines 'usually employed by engineers and machine-makers' – lathes, and planing, shaping and drilling machines—'were generally behind those of England'; while the machinery valuable for its novelty of design was 'roughly constructed (partly of wood) and would not bear comparison in stability and appearance with the highly polished iron machinery of England'.[5] Possibly an element of bias entered into these judgements,[6] and the full significance of the contrast was not analysed.[7]

[1] *Whitworth*, p. 130.

[2] *Rep. on Machinery*, pp. 582 ff. The machines particularized appear to be the precursors of the modern 'automatic' and the 'profiler'. 'The Commission (also) saw much that was new to them which would be valuable to the service in drills for metal, bits for chiselling wood, modes of cutting screws, turning of nipple cones, and in tempering milling tools and cutters.'

[3] Ibid., p. 589.   [4] *Whitworth*, p. 112.   [5] *Rep. on Machinery*, pp. 578, 621.

[6] *Vide infra.* It is of interest to compare with these views those of a French writer, who preferred the American habit of painting metal parts instead of polishing them—the polishing involving additional costs of production and upkeep; while American products exhibited 'la soumission la plus parfaite à la règle générale des constructions; faire les pièces fixes les plus pesantes; les pièces mobiles les plus légères possibles.'—*Revue Universelle des Mines*, etc., Tome XII, p. 295 (1862).

[7] It is only intended in the text to illustrate the character of American

In the years with which this article is concerned there was no other occasion which called forth such elaborate surveys of American engineering developments as those of the early fifties. But there were many incidental judgements, among them one of great interest passed by another of England's most celebrated machine-makers – James Nasmyth – on a factory established in England by Colt employing American automatic machinery for the manufacture of pistols. Colt's factory, established in 1851, did much to form English opinion on the desirability of introducing machine methods for small-arms manufacture. 'It was impossible,' Anderson told a Parliamentary Committee before his American visit, 'to go through that works without coming out a better engineer.' Nasmyth's praise was equally enthusiastic. He confessed himself humiliated by the experience. 'The acquaintance with correct principles has been carried out in a fearless and masterly manner, and they have been pushed to their full extent; and the result is the attainment of perfection and economy such as I have never seen before.' Many English mechanics knew the correct principles, 'but there is a certain degree of timidity resulting from traditional notions, and attachment to old systems, even among the most talented persons, that they keep considerably behind'. Nothing was more impressive than the extent to which unskilled labour could be employed. 'In many cases young men mind four machines. One had been a butcher, another a tailor, another a gentleman's servant. . . . You do not depend on dexterity—all you want is intellect.'[1]

At successive international exhibitions held in Paris and London – in 1856, 1862, and 1867 – the American exhibits, generally very scanty, received the applause of experts, though not in general, of the public. Recognition was not confined to

---

industrial development, and not its scope, as illustrated in the reports. Labour-saving and systematic organization were illustrated over a very wide field. The boot and shoe industry, for instance, provided an example of very systematic outwork; and the suitability of the product was already increased by manufacture in sizes and half-sizes. There were important advances in textiles – especially the development of the carpet loom. Mechanical packing had been begun; packets of cartridge powder automatically weighed were made by machinery, and pins placed mechanically in rows in paper. *Wallis*, pp. 41, 45, 57; *Rep. on Machinery*, p. 608.

[1] *Select Committee on Small-Arms*, P.P., 1854, XVIII, Q. 1367.

English engineers. It is possible to trace in these records some improvement in the standard of workshop machinery, and in the general level of workmanship (though in some branches this had always been high). But there was no change in the main characteristics of American productions, or in the contrasts with English work drawn by the commissioners in the reports which have been analysed.

Meantime machines developed in the United States had been frequently introduced into English industries. Sometimes the machines themselves were imported; more often they were manufactured here to American designs. 'It is the custom among machine-makers in England', John Platt told a Parliamentary committee in 1867, 'to purchase inventions from the Americans and adapt them to use in this country.'[1] The custom did not depend much, it appears, on the English engineers visiting New England – which happened very rarely – but on the initiative of the patentees.

Two instances of American influence are so well known that it is unnecessary to dwell upon them here. The United States provided the chief influence leading to the practical success of the sewing-machine from the early fifties. Large numbers of machines were imported from the States, and a considerable manufacture had been established here by 1860, depending largely on the purchase of American patents. Throughout the years under discussion the main developments in these machines – such as the extension into shoe-making, and the production of devices for buttonholing – were of transatlantic origin with few exceptions.[2] The story of the reaper is similar. Two American machines – those of Hussey and McCormick exhibited in 1851 – provided the stimulus for the manufacture in England; though by the side of these as models there was the Scots machine of Bell. By 1862 the machines had been 'improved and adapted to the circumstances of English crops,

---

[1] *Select Committee on Scientific Instruction*, P.P., 1867–8, XV, Q. 5671. Platt thought he might fairly be described as 'the largest mechanical engineer in the world', employing 5,000 to 7,000 men.

[2] Prof. Willis traced the early history of the sewing-machine in *Reports on Paris Exhibition, 1856*, Vol. XXXVI, Pt. II, p. 168. There were twelve English makers showing at the next Paris exhibition; *Reports on Paris Exhibition*, P.P., 1867–8, XXX, Pt. III, p. 131. Cf. also *Engineering*, Vol. III, p. 513 (1867); and Ure, *Dict. of Manufactures*, 1878, IV, pp. 123 and 130–1, for the application to boot and shoe manufacture.

. . . and there is now no large arable district in the country where the reaping machine is not employed'.[1] It was often an imported machine.

These instances were not isolated, and the scope and importance of American influence can be shown most clearly by tracing some less well-known and individually less important instances. Among these it is perhaps natural to select first the manufacture of small-arms, which was revolutionized as a result of the visits of inspection which have been described. Anderson had introduced into Woolwich before the tours one important American invention – the Blanchard copying lathe for producing irregular shapes to a pattern (such as boot lasts or 'Jacobean' chair legs).[2] The departmental commissioners were given powers to purchase machinery for small-arms manufacture and for the carriage department, and they exercised their powers with that catholicity which marked the selection of factories to visit on their tour. They appear to have purchased a complete set of the machines for the wood and metal work of musket manufacture, including the jigs and gauges, whose 'continual and careful application' was fundamental in securing uniformity and interchangeability; and in addition tin-plate-working machinery, small bench tools – among them being the breast drill with bevel-geared drive – and miscellaneous woodworking machines – for planing, tenoning, mortising, sash-making, and door-making.[3] The United States also supplied the machines – or the models for the machines – in the Birmingham Small-Arms factory which was set up after the model of the government factory at Enfield.[4]

American influence on the use and manufacture of woodworking machinery goes back, however, before 1854 to the Great Exhibition. At this time, although some very ingenious woodworking machines had been devised in England, their scope was far smaller than that of American machines, and

---

[1] *Reports of the Juries, London Exhibition*, 1862, Class IX, pp. 2–3. The growth in output of reaping-machines of four of the biggest firms of agricultural implement makers is given statistically: 1858, 32; 1861, 1,715.

[2] Fairbairn, *Useful Information for Engineers*, 1852, p. 24. (An illustration of a machine of this type is given in J. Rose, *Modern Machine Shop Practice*, 1888, Vol. I, p. 210.)

[3] *Rep. on Machinery*, pp. 621–6.

[4] S. Timmins, *Birmingham and the Midland Hardware District*, 1866, pp. 403–4.

they were not adapted to small workshops. In 1851 a Liverpool firm imported machines from the States and exhibited them; and after some preliminary ridicule on account of the lightness of their construction and the use of wood in their framework, they were acclaimed for their success in operation. Importation continued, and within a few years several English firms had set out on the manufacture of similar machinery. They 'improved' on their models, and their products were more in keeping with English canons of construction, 'far above the more flimsy American products' in Anderson's eyes.[1] But their prices were stated to be twice as high as the Americans'; and some responsible critics disliked the 'needless weight, unnecessary finish and complicated movements' in their work.[2] The States continued to be the home of novel types of machines; 'we are accustomed, it was stated in *Engineering*, to look to America whenever a fresh desideratum in woodworking machinery makes itself felt in general practice';[3] and while perhaps this is exaggerated it is not wholly misleading. In 1867 the biggest circular saw exhibited by a British firm at Paris was of American design,[4] and instances of the adoption of American patents by prominent English makers can easily be traced.[5] The machines seem to have penetrated into all the main branches of woodworking industry, though it is probable that the majority of small workshops had not felt their influence by 1870.

The English hardware industry does not appear to have made much progress in the period discussed here; but where changes occurred the debt to America was big.[6] There were

---

[1] *Whitworth*, p. 116. *Rep. on Paris Exhib.*, 1867–8, XXX, Pt. II, p. 723.

[2] Article quoted from the *Engineer*, in *Engineering Facts and Figures*, 1863, ed. A. B. Brown, pp. 381–4.

[3] *Engineering*, III, p. 665. Also *Artisans Reports*, Exhibition of 1867, p. 220.

[4] Ibid., p. 125. American hand-saws appear to have been much better than English ones. Cf. *Journal of R. Soc. Arts*, 3 September 1875.

[5] *E.g.* in *Engineering*, between 1866–70, prominence is given to the adoption by important English makers of Howe's Veneer Cutter – replacing saw cutters by knives – dovetailing machines by Grover and Armstrong, and improvements to the band saw (itself a French invention). *Engineering*, II, p. 483; III, p. 320; IV, p. 444; VIII, p. 224. Armstrong's Dovetailer is a piquant instance, because, while adopted throughout Europe, it was quickly superseded in America by cheaper machines. Anderson, *Rep. on Philadelphia Exhibition*, p. 1877, XXXIV, p. 323. Cf. also *Reports of Juries, Exhibition of* 1862, Class VII B, p. 19.

[6] This paragraph depends much on *Reports of Juries, Exhibition of 1862*, Class XXXI, particularly pp. 1–3 and 11–12.

three prominent instances of this. The improved methods of wood-screw manufacture – producing for the first time a gimlet-ended screw – were introduced successfully by Nettlefold and Chamberlain in 1854, and were developed with such success that in the late fifties a very large export of screws to the States was created, to be destroyed by the raising of duties from 25 to 75 per cent. In some directions the machines were improved. By 1870 the factory was one of the biggest in the neighbourhood of Birmingham, having about 1,000 employees.[1] A near competitor in size was the factory of the Patent Nut and Bolt Company, set up by an American inventor to develop the mass production (and therefore cheap production) of nuts and bolts of various types, which firms were accustomed to make to meet their own needs. The business had a quick success,[2] and an English firm had been established for the same object by 1862, also using American machinery. The third direction in which American influence was felt was in the manufacture of locks. The employment of automatic machines (some bringing several tools successively into operation) in lock-making was introduced into England from the States by W. Hobbs in 1851. His London factory product had lowered the price of cheap lever locks by the early sixties, but not enough to induce the general adoption of his methods; and in 1870 factory reports treat his works as the only one employing machinery 'like that used at Enfield'. The reason for this must be found partly in the expense of tools where a single article was needed in a great variety of sorts and sizes, partly in the extreme cheapness of handicraft work in the industry.[3] It is curious that clockmaking, the third industry in America singled out by the commissioners as producing metal mechanisms with interchangeable parts, was in England unaffected by the new methods; and this applied also to watchmaking, though severe American competition in this industry had been established by 1865.[4]

That the United States had made important contributions to workshop machinery has been already abundantly illustrated.

[1] *Engineering*, VII, p. 385, and Timmins, op. cit., pp. 605–8.

[2] Ibid., VI, p. 534. The capital in 1867 was £400,000.

[3] *Report of Juries*, 1862, loc. cit., Timmins, op. cit., p. 90; *Children's Employment Commission*, 3rd Rep., P.P., 1864, XXII, pp. 7 and 25–28.

[4] *Horological Journal*, 1867, pp. 86–7.

But the contribution was not restricted to the types of machinery so far described. American planing and screwing machinery was employed extensively in England in the sixties;[1] by 1865 the Morse Twist Drill Company was manufacturing drills in graded sizes for the trade, thus 'doing for drilling instruments what Sir Joseph Whitworth did for screwing apparatus';[2] and by the same time the Brown and Sharpe micrometer calliper was being manufactured, though probably it was not used here.[3]

Another industry in which England had been the recent and undisputed pioneer – the steam-engine manufacture – owed much in these years to American ingenuity, the chief indebtedness arising from the adoption of the Corliss engine. Its peculiarity lay in its valve gear, and with modifications it was introduced here first in 1862. Within a few years it had established a reputation both for fuel economy and for uniformity of speed under a variable load. The Americans, it was said, when faced with relatively expensive coal, 'made fuel out of brains'. By 1867 at least fifty Corliss engines had been manufactured here, three of them for Woolwich Arsenal and one for John Platt; and there were seven firms manufacturing them. One of these firms – Messrs Hicks, Hargreaves & Co. – is said to have produced 1,400 of these engines since that date.[4] An American designer, Allen, was also responsible for the introduction into England of the principle of the quickly-moving piston; his engine was employed by Whitworth, and with modifications

[1] The machines freely used were those of Sellers, whose display was said by Anderson to have 'contained more originality than that of any other exhibitor' of similar articles at Paris in 1867. *Rep. on Paris Exhib.*, P.P., 1867–8, XXX, II, p. 716. (The exhibit is described.) Sellers first patented a planing machine in England in 1860, and his bolt-screwing machine, dispensing with any return motion, was shown by an English firm in 1862. *Reports of Juries*, 1862, Class VII B, p. 12.

[2] *English Mechanic*, 1865, p. 77. *Rep. on Philadelphia Exhib.*, P.P., 1877, XXXIV, p. 320.

[3] *Engineering*, VIII, p. 401.

[4] Ibid., II, pp. 3, 98–9; III, p. 658; IV, pp. 8, 155; *Rep. on Paris Exhib.*, P.P., 1867–8, Pt. II, p. 649. *Journal of the Textile Industry*, Special (Crompton Centenary) Issue, July 1927, p. 111. The Corliss engine was also made in Germany and Switzerland, and was made in England without English modifications by Galloway & Co. It was adapted as a marine engine (*Engineering*, V, p. 471). In the United States it was made with standardized parts, which was not the practice with steam-engine manufacture in England (*Artisan*, 1863, p. 159).

D

by another designer from the States was manufactured by Whitworth for the English market.[1]

To show more amply, though with no aim at completeness, the scope of the American export of inventions, four final examples from unrelated industries may be given. Gwynne, whose centrifugal pumps were much commended at the 1851 exhibition, established a factory here which rapidly obtained an extensive trade.[2] It is probable that the stimulus for the introduction of machine methods into brick-making came largely from the United States; Whitworth certainly advocated it as a result of his tour.[3] Whether this be the case or no, one of the machines which achieved marked success in the sixties was of American origin. Most of the successful machines used wet clay, and the resulting bricks were of poor quality in many cases; the patent which was bought by Platt was a 'dry-clay' machine; its product was of good quality, employed by railway companies; and the machine was sold both in England and for export.[4] Stone-breaking by machine appears to have waited on American invention; Blake's patent machine was brought into England in 1862 by H. R. Marsden, and employed both for road work and for ore-crushing; and it remained prominent.[5] Finally, it is of some interest that *The Times* and the leading Manchester newspapers were printed in the sixties by Hoe presses, designed by what was then a well-known New York and Boston firm, and made by Whitworth. In 1869 the American company opened its London factory.[6]

Within the years discussed in this article, years when by textbook tradition England was the workshop of the world, influences of the type chronicled here formed the main influence of the United States' industry on England; and they constituted a remarkable contribution to the 'retreat of the handicrafts'. Some branches of English industry were, however, otherwise – and adversely – affected. It became less possible to

[1] *Engineering*, V, p. 119; *Rep. on Paris Exhib.*, P.P., 1867–8, Pt. II, p. 647. There were many other U.S. influences on steam-engine practice which it is impossible to trace here.

[2] B. P. Johnson, *Report of N.Y. Commissioner to Exhibition*, 1851, p. 67.

[3] *Proc. Inst. Mech. Eng.*, 1856, p. 128.

[4] *Engineering*, III, p. 197.

[5] Ibid., IV, p. 167. W. H. Maxwell, *British Progress in Municipal Engineering*, 1904, p. 41.

[6] *Engineering*, VII, p. 235.

say, what had been (rather clumsily) said in 1851, that there was 'no satisfactory proof of our industry being beaten in perfectly neutral markets, or at present any signs of its being likely to'.[1]

Probably this situation had developed most in regard to certain Birmingham industries, and it was described in 1867 to a Parliamentary Committee by A. Field, President of the Birmingham Chamber of Commerce, who was concerned in the hardware export trade. It is of interest that there was not only a much diminished consumption of Birmingham hardware in the United States, but continental competitors were succeeding where we failed, and in face of the same tariff. It is more apposite here, however, that Birmingham was meeting successful competition from the United States in South America and the colonies – Australia as well as Canada – in such things as ploughs, shovels, nails, tools, pumps, locks, and petroleum lamps. Field ascribed the success of competition to the qualities of the American product.

> English goods in my line are not altered to meet new requirements . . . the United States' goods are rapidly changed. . . . The American sees the article will answer the purpose it is made for . . . he frequently omits a considerable portion of that which has been considered important to the construction of the article in England, and he contrives by some direct method to make it efficient for its purpose, and generally more efficient than the Englishman makes it; it is often more simple in construction, and always lighter in weight.[2]

The process described by Field can be illustrated to some extent by the foreign trade statistics of the United States. In Table I the values of some of the main exports of manufactures of iron and steel in 1867–8 are given, with their chief markets distinguished. The trade had grown rapidly since the early fifties; but owing to continual changes in classifications of

[1] *Exhibition Lectures*, 1851; Second Series, p. 441.

[2] *Report of Select Committee on Scientific Instruction*, P.P., 1867–8, XV, Q. 6720–6785. On p. 513 of this report an extensive list is given of articles where the produce of the United States, and to a less extent Germany, France, and Belgium, have 'replaced those of Birmingham and the Midland Hardware District in whole or in part'.

exports of manufactures, the rate of growth cannot be estimated with precision. It is clear that, in spite of the Civil War, the total value of iron and steel manufactures exported more than doubled between 1851–2 and 1867–8, after allowances have been made for the depreciation of the dollar; and the proportion of these exports composed of mechanisms of various kinds must have grown considerably more rapidly.

TABLE I *Representative Exports of U.S. Iron and Steel Manufactures, 1867–8; with their chief destinations*[1]
(Values in 1,000 dollars)

Country	Agricultural implements	Locomotives and other machines	Clocks	Sewing-machines	Unspecified iron manufactures	Cutlery	Muskets and rifles
England	35·1	264·9	262·5	621·8	115·0	13·9	917·6
Brit. N. America	92·2	103·0	23·5	27·2	39·8	1·3	0·3
Brit. Africa	50·5	11·3		3·5	10·3		0·1
Australia	56·5	14·6	30·3	57·8	298·7	65·3	1·3
France		84·3		107·8	6·4	0·3	17·1
Hamburg	0·8	130·0	48·1	377·0	12·2	7·4	184·1
Bremen	2·5	8·5	2·6	0·7	3·9	1·0	257·9
Argentine	114·6	28·4	26·4	46·4	124·3	16·9	0·1
Brazil	51·0	133·5	17·5	125·1	97·0	35·1	135·5
Chile	67·2	114·1	2·3	15·7	105·9		16·7
Colombia	22·4	463·4	0·4	23·0	393·3	10·3	45·9
Cuba	91·2	704·0	18·5	49·0	473·2	49·0	
Mexico	21·4	331·1	4·3	26·9	243·9	15·8	42·5
Peru	16·9	71·1	2·2	23·0	50·2		360·0
All Countries	673·4	2,577·0	536·7	1,658·0	2,812·1	221·5	2,483·8

It is useful to place beside these figures some parallel statistics of English exports. In Table II the total values of English exports of machinery, agricultural implements (including machinery, save steam engines, for agricultural use), hardware and cutlery, are given for 1867, with the values which were destined for Australia, the North American colonies, and four South American states. It will be seen that the values of American exports of machinery (including sewing-machines) and the combined 'Iron Manufactures' and 'Cutlery' were

[1] U.S. Executive Documents: Trade and Navigation.

equivalent respectively to 11 per cent and 12 per cent of the values of comparable classes of English exports; in the case of Agricultural Implements the figure was 25 per cent. But the proportionate importance of the United States' exports in relation to the English exports for the countries for which details are given in the following table was much above these figures, save in the case of Australia. British North America

TABLE 11 *Exports of Machinery, Agricultural Implements, and Hardware and Cutlery from Great Britain, 1867*[1]

(Values in £1,000)

Country	Machinery	Agricultural implements and machinery	Hardware and cutlery
British North America	20·4	4·4	259·3
Australia	186·2	73·2	286·5
Argentine	28·0	2·5	120·3
Brazil	56·1	13·9	163·8
Chile	67·5	3·9	88·6
Peru	123·0	3·8	43·3
All Countries	4,968·5	381·2	3,941·6

and the South American states took machinery to the value of at least 35 per cent of the corresponding imports from England; while all save Brazil imported agricultural implements and machinery from the United States of considerably greater value than came from England. The generalized statistics naturally

[1] Compiled from the *Annual Statement of Trade and Navigation*, P.P., 1868, LXVII. The figures in the first column include the values of all machinery exported save agricultural machinery: 'Agricultural implements and machinery' include items listed under 'Agricultural Implements', 'Iron Manufactures for use in Agriculture', and 'Agricultural Machines'. The basis of the figures in the third column is explained on p. 144 of the Annual Statement. In a comparison of these figures with those in Table I it is to be remembered that the dollar was depreciated at this date, and it is roughly justifiable to take £1 as equivalent to $7. While not strictly apposite to this article, it is interesting that the continental competition with English products in the American market alluded to by Field is traceable to some extent in the American statistics. One element of it may be shown briefly in tabular form.

Imports of Cutlery into United States (dollars)

	1860	1866	1867	1868	1869
From Great Britain	2,070,267	2,270,306	2,033,430	1,361,116	1,358,765
From Elsewhere	169,618	284,896	363,121	216,499	137,143

only illustrate the position rather clumsily; and it is doubtful whether from existing statistics it would be possible to compare the relative growths of comparable English and American exports of iron and steel manufactures with these markets in the fifties and sixties.

Those who observed American advance in machine-making and machine-using turned naturally to try and explain its origins; and their not very systematic analysis is composed mainly of familiar features, though the conditions governing the supply of raw materials were in many respects, of course, sharply contrasted with the present position.

All explanations lay considerable emphasis on the peculiarities of the market, particularly on its size. The size of the population, growing by emigration as well as by natural increase, was such that 'whether the supply of goods is derived from the home or the foreign manufacturer the demand cannot fail to be greater than the supply'.[1] The opportunity afforded by this was increased by the higher average wealth of the people. Wallis noticed how greatly the demand for ready-made clothing was added to by the fact that 'all classes of the people may be said to be well dressed and the cast-off clothes of one class are never worn by another'.[2] In some respects this market was also extremely receptive; no doubt due in part to inadequate supply. Of the hardware trade it was said that

> the willingness on the part of the American public to buy what is offered them, if it can in any way answer the purpose, has given a great advantage to the North American manufacturer over his European competitor, who has to contend with habits and prejudices of centuries standing, and even now almost in full vigour. . . . In the United States they overlook defects more than in Europe, and are satisfied if a machine intended to supersede domestic labour will work even imperfectly, while we insist on its being thoroughly well made and efficient.

---

[1] *Wallis*, p. 78. 'This once fairly recognized,' runs the free-trade corollary, 'those jealousies which have ever tended to retard the progress of nations in the peaceful arts, will be no longer suffered to interfere by taking the form of restrictions on commerce and the free intercourse of peoples.'

[2] *Wallis*, p. 43.

In England the public had 'fixed notions of shapes, sizes and prices', and demand increased slowly enough to be supplied 'in the old way, by manufacturers using skilled labour'. The market of conventional consumers would not facilitate a transition to new methods, in the early years when the new methods did not greatly lower prices.[1]

The English observers also stressed the relation between the characteristics of American industry and the peculiar difficulties which had to be overcome in order to establish it successfully. There was, in the first place, the relative dearness of some of the primary raw materials. The alleged influence of dear coal on steam-engine design has been referred to earlier.[2] The high price of iron was held partly responsible for the lightness of structure in machines,[3] which was often adversely criticized in England, but which has had a lasting influence on design. (Probably it was not merely a coincidence that Whitworth within three years of his tour was advocating greater lightness in the moving parts of mechanisms.)[4] The expense of raw materials was due partly to the high price of labour, and thus reflected the second difficulty attending American industrial advance. Among skilled artisans expense was due to scarcity; among the unskilled to the opportunity of settlement on the land. The position was regarded not merely as a constant inducement to the adoption of labour-saving devices, but as a stimulus to the inventive faculty.[5] Wallis also laid emphasis on the influence in certain trades – probably mainly small metal industries – of difficulties which faced the immigrant skilled artisan in carrying on his industry, owing to the absence of the customary environment which had grown up in Europe. There

[1] *Reports of Juries*, 1862, pp. 2–3. Apart from these general conditions of demand, the existence of special demands peculiar in intensity through special circumstances was recognized – e.g. forest clearing encouraged perfection of axe manufacture. It is scarcely necessary to add that while the general contrast of English and American demand was sound, intense demands existed in England – in the building trade, for instance – without the response in new method which the rather facile popular use of the large-market argument expects.

[2] *Vide supra.*

[3] This influence is stressed in a favourable account of an American lathe in *Reports of the Juries, Exhibition of* 1851, p. 199.

[4] *Proc. Inst. Mech. Engineers*, 1856, p. 126. Whitworth was also influenced by Bessemer's process.

[5] English writers constantly referred to this: e.g. *Wallis*, pp. 12, 13; *Whitworth*, p. 41; *Rep. on Machinery*, p. 578; *Report of Juries*, 1851, p. 201.

he had been accustomed to the existence of persons carrying out complementary work; obtaining the raw material and performing the early stages of manufacture. In the new environment it was necessary to organize all stages of manufacture, hence a tendency to the integration of processes, and the adoption of a systematic factory structure. Wallis also regarded this factor as an important impulse to the adoption of machine methods.[1]

While difficulties such as these placed the American manufacturer in the beginning at a disadvantage, there was one matter in which the Commissions of the early fifties found he was placed in a more fortunate position for developing factory industry than the English producer. They discovered that most of the factories were owned by corporations in which 'the liabilities of partners not actively engaged in the management are limited to the proportion of the capital subscribed by each'. Many such companies had been formed by incorporation even before 1800, and laws facilitating their formation had been introduced in several states in the early years of industrial growth during the first quarter of the nineteenth century. The expenses of forming such corporations were negligible, and it was obvious that the limit of risk encouraged enterprise and attracted capital from sources, many of them small, which would otherwise have remained untouched, perhaps undeveloped through the lack of sufficient facilities for investment. Though Wallis was careful to state that he saw grave dangers of inefficient management in these companies, he agreed that the laws encouraging their formation had 'led to a much greater development of the industrial resources and skill of the country than, in its circumstances, could have resulted from mere private enterprise for many years to come'. His evidence did not in the least illustrate the dangers of which he spoke, and the general effect of the reports must have been to add considerably to the force of the movement for a change in the English company law.[2]

---

[1] *Wallis*, p. 12.

[2] *Rep. on Machinery*, p. 578; *Wallis*, p. 14; *Whitworth*, p. 113. The importance of the American form of company had been remarked by earlier English writers: e.g. Wilson, *Partnerships en Commandite*, 1848, *passim*. Some particulars (which might be more precise) about the history of the company laws of the United States are given by Clark, op. cit., Vol. I, pp. 266–8.

Factors of the kind discussed might affect the direction taken by American developments, but the dynamic factor, it was recognized, was to be found in the character of the people – in their energy, enterprise, and ingenuity. These qualities were widespread among all grades engaged in industry; and their presence among the artisans was frequently emphasized. According to Wallis, 'traditional methods had little hold upon the American as compared with the English artisan, and processes holding out the least promise of improvement were quickly tested'.[1] The Ordnance Commissioners found that masters and men were convinced that labour-saving devices were for their mutual benefit, and that 'every workman seems to be continually devising some new thing to assist him in his work.' They were all anxious to be 'posted up' in every new development.[2] Whitworth gave the same testimony.

In some measure these facts could be explained by the partial recruitment of industrial labour by immigration. Immigrants from England, Germany and France brought the varied traditions of the leading industrial nations together; and it was possible to adopt the best elements of what had been conflicting practices. The reports of the early fifties illustrated one aspect of this clearly by ascribing the superiority of American castings over English to German influence.[3] It is not improbable that immigration brought into the country persons who were enterprising beyond the average; and it was frequently remarked that once the tide of industrial advance had set in in the States, America became the Mecca of the European inventor.[4]

A partial explanation of the energy and enterprise exhibited in American industry was also sought, in another direction, in two conditions of American industrial life which were sharply contrasted with English conditions – the absence of a fear of unemployment and the mobility between industrial employments. Whitworth explained the absence of trade unions and their restrictions on the introduction of new methods (which many English business men noticed enviously) as largely due

[1] *Wallis*, p. 15.
[2] *Rep. on Machinery*, pp. 578, 584.
[3] *Wallis*, pp. 16, 53; *Rep. on Machinery*, p. 604.
[4] *Rep. of Juries*, 1862, Class XXXI, p. 3; Lovett, *Life*, Vol. I, p. 65 (Bahn's Popular Edition).

to the relative scarcity of labour. 'With a superabundant supply
of hands in this country (i.e. in England), and therefore a proportionate difficulty in obtaining remunerative employment,
the working classes have less sympathy with the progress of
invention.'[1] In some respects the mobility between employments, the frequency with which men moved not only from
one branch of an industry to another, but from one industry or
occupation to another,[2] might be regarded mainly as a result
of the peculiar American conditions; of the growing market
and opportunities, and the enterprise shown in taking openings.
But it was also a cause of enterprise and improvement; partly
because it averted boredom,[3] partly, one may assume, because
it helped in the circulation of ideas, since methods would
spread rapidly in related trades where there were similar technical problems; and partly because the fact of extensive and as
it seemed successful change provoked a straining after change,
set up a fashion or ideal. 'The American was a believer to an
unlimited extent in progress', as Fraser said in his valuable
survey of American education in 1867. 'Universal movement,
acting on natures peculiarly susceptible of its influence',
explained to a large extent, in his view, the intelligence and
versatility of the people;[4] and while obviously this influence
could not explain the beginnings of American advance, it was
most likely important for its persistence.

Fraser was inclined to minimize the influence of the 'only
partially excellent schools', which he had surveyed elaborately.
He seems to suggest that the schools were much more a reflection of characteristics than a cause. In this he was mini-
mizing an influence which several of the writers whose views
have been discussed in this article rated very highly.[5] Wallis
looked upon the 'sound practical education' which was common in America as fundamental in its industrial growth. The
remarkable application of machinery displayed 'the adaptive

---

[1] *Whitworth*, p. 146.

[2] *Wallis*, p. 13.

[3] Ibid.

[4] J. Fraser, *Report to School Inquiry Commission*, P.P., 1867, XXVI, p. 496.

[5] Fraser did not deny all influence to the education, but held that it was more
liable to supply information than to develop mental power, and that the scientific
education given was superficial. It was, however, a channel through which a
desire for change and a desire to excel were inculcated. *Report*, pp. 469, 476, 496.

versatility of an educated people'. Particularly in the New England states, where the industrial movement had its centre, 'the skilled workmen . . . are educated alike in the simplest elements of knowledge as in the most skilful application of their ingenuity to the useful arts and the manufacturing industry of their country'; and with their 'perceptive power so keenly awakened by early intellectual training' they quickly learned from other nations, and improved upon them.[1] Whitworth similarly stressed the importance of education, and made it the subject of his peroration. He noted that it never happened in the States, as it did in England, that a man with the necessary technical skill was unable to take up a position as foreman through a lack of the primary elements of education; and he associated with education in importance the very extensive Press. 'Where the humblest labourer can indulge in the luxury of a daily paper, everyone reads, and thought and intelligence penetrate through the lowest grades of society.' The benefits of liberal education and a cheap Press in the United States 'could hardly be over-estimated'.

> Wherever education and an unrestricted Press are allowed full scope to exercise their united influence [he wrote in his concluding paragraph] progress and improvement are the certain result; and among the many benefits which arise from the joint co-operation may be ranked most prominently the value they teach men to place upon intelligent contrivance; the readiness with which they cause new contrivances to be received, and the impulse which they give to that inventive spirit which is gradually emancipating man from the rude forms of labour, and making what were regarded as the luxuries of one age to be looked upon in the next as the ordinary and necessary condition of human labour.[2]

At the end of the period considered here the same view of the importance of education in the States was repeated by Field in the evidence which has already been partially quoted. 'The cause of the whole difference between us is to be found in the education on the two sides of the water,' he declared; and he stated the view at some length. Among other advantages he

[1] *Wallis*, pp. 13, 77. [2] *Whitworth*, pp. 146–7.

noticed that 'trade unions and strikes cannot exist in that educated atmosphere'.[1]

It is probable that Whitworth and Wallis were influenced in their treatment of American education by the potential effect of their reports in matters of industrial politics. Among factors whose relative importance was difficult to weigh it would be natural to select for most emphasis those in which England's disadvantage seemed most readily removable. The commissioners of the 1851 Exhibition had already recently drawn attention to the part played by education in the emergence of continental competition. Those countries who lacked cheap fuel or raw materials tended 'to depend more on the intellectual elements of production than in this country'. The Reports of 1854 gave valuable support to the people who were emphasizing the industrial importance of education, though it needed the influence of three more international exhibitions to evoke a movement of real political significance.

[1] *Sel. Comm. on Scientific Instruction*, P.P., 1867–8, Q. 6723, 6763.

3 The Enfield Arsenal in Theory and History[1]

EDWARD AMES and NATHAN ROSENBERG

[This article was first published in the *Economic Journal*, Vol. LXXVIII (December 1968).]

There is now a substantial body of historical literature concerning nineteenth-century development and technical change.[2] This literature relies, of course, upon detailed discussion of events in particular industries, but its objective is in some sense macro-economic. That is, it aims at making comparisons of levels and rates of change in techniques for groups of industries or even countries. This objective requires the aggregation of

[1] An earlier version of this paper was presented at Professor Robert Fogel's workshop in Economic History at the University of Chicago. We are indebted to the participants, whose puzzlement at some of our analysis forced us to several reluctant clarifications.

[2] E. Rothbarth, 'Causes of the Superior Efficiency of U.S.A. Industry as Compared with British Industry', *Economic Journal*, September 1946, pp. 383–90; D. L. Burn, 'The Genesis of American Engineering Competition, 1850–1870', *Economic History*, January 1931, pp. 292–311; John Sawyer, 'The Social Basis of the American System of Manufacturing,' *Journal of Economic History*, December 1954, pp. 361–79; H. J. Habakkuk, *American and British Technology in the Nineteenth Century* (Cambridge, 1962); Peter Temin, 'Labor Scarcity and the Problem of American Industrial Efficiency in the 1850s', *Journal of Economic History*, September 1966, pp. 277–98. Temin has drawn heavily upon the accounts of British Parliamentary Commissions which visited the United States in 1853 and 1854. The reports of these commissions provide the most authoritative account by contemporaries of Anglo-American technological differences in the middle of the nineteenth century. Moreover, the second commission was directly responsible for the introduction into England of American gunmaking machinery. During its tour of the United States the Committee on the Machinery of the United States purchased large quantities of specialized gunmaking machinery which formed the basis for the expanded Enfield Arsenal. See *New York Industrial Exhibition: Special Report of Mr Joseph Whitworth and Special Report of George Wallis*, P.P., 1854, XXXVI; *Report of the Committee on the Machinery of the United States of America*, P.P., 1854–55, L. The hearings of the Small Arms Committee in March and April of 1854 are an indispensable source of information on the British firearms industry and on the history of the Government's parlous relationship with that industry. See *Report from the Select Committee on Small Arms*, P.P., 1854, XVIII.

results about individual industrial operations. All aggregation suppresses information about the variables being combined, and it is important to inquire about the consequences of aggregation in analysis of this particular sort.

This paper analyses a particular historical event, the establishment of the Enfield Arsenal, in the context of the literature cited. The British Government committed itself to the construction of the Enfield Arsenal in 1854 because it wished to be able to make large numbers of rifles for an impending war with Russia (now known as the Crimean War). The event is important because it marked the beginning of the movement of mass-production techniques from the United States to Europe. Technical changes in gunmaking in the nineteenth century were a major source of new machine techniques; and industrialization in the nineteenth century is overwhelmingly the history of the spread of machine making and machine using.

Before 1854, British gunmaking was concentrated in a large but complicated structure of handicraft firms, mainly located in Birmingham, and producing firearms to individual order or in very small batches. American gunmakers were at this time engaged in mass production of both civilian and military weapons. These weapons had interchangeable parts, a fact which the British found to be almost unbelievable. This production required a number of machines which were virtually unknown in Britain before the hearings. The Enfield Arsenal was equipped almost entirely with machinery of American design and manufacture; and American workers were brought to England to introduce the machines and to train English workers in their use.[1]

The Parliamentary hearings of 1853–4 and other sources stress the differences between what we shall call the Birmingham and Enfield processes. In order to avoid being overwhelmed by the details on gunmaking, we shall analyse technical change in gunmaking according to the precepts of the theory of the firm.

The central question considered is the following: is the

[1] The full details concerning the establishment of the Enfield Arsenal with American equipment are related in *Great Britain and the American System of Manufacturing*, edited with an introductory essay by Nathan Rosenberg (Edinburgh University Press, 1968).

difference between the English and American gunmaking industries to be regarded as a difference in technology, or merely as a response to the differing price structures in the two countries? On the principle of Occam's razor, economists should hesitate to introduce a somewhat nebulous concept, 'technology', into a phenomenon which might be explained by market analysis. Consequently, they should be attracted to the following hypothesis of Temin's.

Imagine that industrial processes involve 'labour' and 'capital', with a production function $Q \leq g(L, K)$. Under conditions of pure competition, the equilibrium level of output is at the lowest point on average cost curves. At this lowest point there are constant returns to scale, so that the production function is 'homogeneous of degree 1'. It may then be shown that the capital–labour ratio depends on the interest rate (i.e. is independent of the wage level). We know that the interest rate in nineteenth-century America is higher than in nineteenth-century England, and therefore Temin would predict a lower capital–labour ratio in America than in England. He also would predict that the capital-intensive 'Enfield' process should have originated in England.

Finally, he would predict that British production processes would be more capital-intensive than the American techniques.

None of these predictions is borne out by the record. Temin does not reject the theory, however; he rejects the measures of capital intensity, saying that American machinery was more 'flimsy' than the British machinery and is overestimated by the statistics.[1] He also argues that the English knew about the Enfield process before 1854, but did not use it because of the differences in price structures.

We, however, shall argue that it is more appropriate to look for a new theory. In explaining our motives for seeking such a theory we shall take advantage of the neat reasoning involved in Temin's demonstration, but apply it to a somewhat different purpose. We therefore reconstruct his argument in a little more detail.

Temin assumes that there are three 'factors of production', land (T), labour (L), and capital (K), and two outputs, agriculture (A) and manufacturing (M). Two outputs mean two

[1] Temin, op. cit., p. 291.

production processes; to keep everything simple, Temin
assumes that no capital is used in agriculture and no land is
used in manufacturing. Thus his production processes are

$$Q_A \le f(L, T)$$
$$Q_M \le g(L, K)$$

Since there is pure competition and constant returns to scale,
we may rewrite

$$Q_A \le LF(T/L) \equiv LF(t)$$
$$Q_M \le LG(K/L) \equiv LG(k)$$

and as Temin shows, under these conditions, the capital–
labour ratio depends only on the interest rate, and the land–
labour ratio depends only on the price of land.

Let us apply this argument to gunmaking. We shall carefully
re-label the symbols. Let there now be three inputs, common
labour (L), craftsmen (K), and machinists (T), and two kinds of
output, Enfield guns (A) and Birmingham guns (M). In the
short run, plant is fixed. We assume that there is pure competition, that Birmingham uses no machinists and Enfield uses no
craftsmen.[1] Then Temin's theorem tells us that the ratio of
machinists to common labour in Enfield plants depends only
on the wage of machinists, and the ratio of craftsmen to
common labour in Birmingham plants depends only on the
wages of craftsmen.

Observe at this point that the concept of a homogeneous
labour force has collapsed. Using a logical structure identical
with that of Temin, we can explain why the composition of
the labour force will depend on the wage structure, so that
(given 'capital') output does not meaningfully depend on the
total labour force. The logic, of course, yields an interesting
suggestion to historians, which could not have been made if
labour were treated as homogeneous.

Suppose that in 1850 both processes had been known in
England and also in America. Temin's theorem predicts when
England would have used mainly the Birmingham process
and America mainly the Enfield process: if the wages of common labour were higher (relative to gun prices) in America
than in England, America will use less common labour than

[1] 'Enfield' is shorthand for 'guns made with interchangeable parts by American methods'.

England. If machinists are relatively plentiful in America (compared to England) their wages will be relatively low in America. If craftsmen are relatively plentiful in England (compared to America) their wages will be relatively low in England. Thus Birmingham methods will predominate in England, and Enfield will predominate in America.[1]

There is fragmentary evidence that, as early as the 1820s, the wages of machinists were relatively low, and the wages of other skilled labour were relatively high in America, as compared to England. Such pay differentials may well have contributed to the differences between English and American gunmaking, and perhaps other machine-making industries.[2]

Of course, if capital is not homogeneous, but consists of craftsmen's tools and machine tools, then even if the interest rate determines the ratio of tools to workers in both the Birmingham and Enfield processes, it is not possible to make meaningful comparisons of 'capital–labour ratios' in the two systems of production. The numerators of the ratios are measured, respectively, by the amount of craftsmen's tools, and the amount of machine tools.[3] If the two countries have different price structures there is no unambiguous way to combine the two kinds of 'capital'.

Our conclusion is that two-input (labour and capital) models are inadequate in discussing nineteenth-century gun-

---

[1] In this connection two fragmentary pieces of evidence are of interest.

(*a*) A number of American machinists had to be imported and employed at the Enfield Arsenal when specialized gunmaking machinery was introduced and had to be adapted to the production of British military firearms. See John Anderson, 'On the Application of Machinery in the War Department', *Journal of the Society of Arts*, 30 January 1857, p. 156, and *The Engineer*, 11 January 1856, p. 20.

(*b*) The American armouries, earlier in the century, were apparently heavily dependent upon foreign (primarily English) sources of gunmaking craftsmen. Charles Fitch ('Report on the Manufacture of Interchangeable Mechanism', *Tenth Census of the United States*, 1880, Vol. II, p. 5) stated that, in 1819, most of the filers at the national armouries were foreign. Another source, apparently referring to the mid-1820s, states that there were over fifty British craftsmen employed at Springfield Armory and at least twenty-one at Harper's Ferry (see handwritten remarks of George Lovell, Inspector of Small Arms, inserted into copy of *Observations on the Manufacture of Fire-Arms for Military Purposes* (London, 1829), at British Museum).

[2] See Nathan Rosenberg, 'Anglo-American Wage Differences in the 1820s', *Journal of Economic History*, June 1967.

[3] Temin, op. cit., p. 291.

making. Three-input theories are the staple fare of Anglo-American economics, and it is natural to consider one application of such theories. Suppose pure competition, as before, and that gunmaking uses 'land' as well as labour and capital. Then the production process may be written

$$Q \le f(L, K, T) = LF(K/T, T/L) = LF(k, t)\begin{cases} k = K/L \\ t = T/L \end{cases}$$

since it is homogeneous of degree one. Now there are two ratios, the capital–labour ratio (k) and the land–labour ratio (t). The optimal values of these ratios depend both on the price of capital and on the price of land. 'Land' in the context of nineteenth-century gunmaking has little to do with agriculture. But guns are made of iron, except for their wooden stocks. The iron comes from ore, and is smelted with either charcoal or coke. Ore, coal, and timber resources are clearly 'land' in the context of gunmaking. America in the nineteenth century was considered (both by Americans and Europeans) to be a centre of 'inexhaustible' natural wealth. How does this fact affect gunmaking? One example will suffice, and will suggest the high cost of compression in moving from a three-factor to a two-factor model.

There is evidence which suggests that the woodworking machines which were popular in America and neglected in England were not only labour-saving but also wasteful of wood. Their adoption in America and neglect in England may be attributable not only – or perhaps not even primarily – to differences in the capital–labour ratios in the two countries but rather to the cheapness of wood in the United States and its high price in England.[1] If nineteenth-century machine processes

---

[1] *Report of the Small Arms Committee*, op. cit., Q. 7273–81 and Q. 7520–21; G. L. Molesworth, 'On the Conversion of Wood by Machinery', *Proceedings of the Institution of Civil Engineers*, Vol. XVII, pp. 22, 45–6. In the discussion which followed Mr Moleworth's paper Mr Worssam, a prominent English dealer in woodworking machinery, made some interesting comparative observations which were summarized as follows: 'He had seen American machines in operation, and he found that, although they might be adapted for the description of work required in that country, they were not so suitable for English work, in which latter high finish and economy of material were of greatest importance. In America the saws were much thicker than those used in the English saw-mills, so that they consumed more power, wasted more material, and did not cut so clean, or so true, though there was less care required in working them' (ibid., pp. 45–6).

were generally more wasteful of raw materials than the corresponding handicraft processes this fact would help to explain the longer persistence of handicraft methods in England than in the United States. We suggest that this may have been the case in woodworking machinery, where, by general agreement among all observers and commentators, the technological differences between England and the United States were the most remarkable. Analysis of this historical problem in terms of a two-factor, capital-and-labour model, therefore, is fundamentally inadequate because such a model is incapable of calling attention to the importance of the price of 'land', which may have been of major importance in the nineteenth century.

The discussion thus far shows that there is evidence in favour of a theory which says that labour is meaningfully heterogeneous; it suggests further that if 'land' is recognized as an input some of the difficulties about explaining international differences in gunmaking practices might be dealt with. These various results all point to the recognition that production processes use many kinds of inputs. Failure to utilize this piece of information has led, in the problem under study, to wrong predictions, and to the supression as meaningless, of important parts of the available record. Suppose, therefore, that we start again, in the light of the theory of the firm.

Firms and industries may be thought of as having two kinds of interrelated problems: technical and market problems. Technical problems are summarized by what the theorist calls a production function. The production function specifies the maximum rate at which the firm or firms in question may produce goods given each possible rate of resource utilization. Formally, the theorist writes $y \leq f(x_1, \ldots x_n)$, where y stands for the rate of output per unit of time and f the maximum which a firm can produce (employing a specific technique), using inputs $1, \ldots n$ (these numbers stand for the *names* of the resources used) at rates $x_1, \ldots x_n$ per unit of time. An inequality appears, for the firm can always operate inefficiently (i.e., produce less than the optimum denoted by f). When economic (market) data are added the firm has the information it needs to decide on a rate of output, and on rates of input use. But we will first consider production functions in the context of historical problems.

Historians often have to consider real production processes. They must, for instance, replace the numbers 1, ... *n* by the names of inputs actually used in a given situation. It makes a difference, for example, whether one of these numbers stands for a hand chisel or a Blanchard gunstocking lathe.

The gunstock was one of the most serious bottlenecks in firearms production. In England, at the time of the Parliamentary hearings, out of about 7,300 workmen in the Birmingham gun trade, the number of men employed in making gunstocks totalled perhaps as many as 2,000.[1] Its highly irregular shape for long seemed to defy mechanical assistance, and the hand-shaping of the stock was a very tedious operation. Furthermore, the fitting and recessing of the stock so that it would properly accommodate the lock and barrel were extremely time-consuming processes, the proper performance of which required considerable experience. With Birmingham methods, it required 75 men to produce 100 stocks per day. Using the early (1818) version of the Blanchard lathe, 17 men could produce 100 stocks per day.[2] Before the Blanchard lathe was introduced, the production function had as arguments (say) r kinds of labour and s kinds of machines: $y \leq f(l_1, \ldots l_r, m_1, \ldots m_s)$. After the introduction of the lathe there is a new production function: $y \leq g(l_1, \ldots l_r, m_1, \ldots m_s, m_{s+1})$. There are now $(s + 1)$ kinds of machines which contribute to output.

Even though a production function has many attributes other than the list of the names of the inputs it uses, the record shows that English engineers found much of interest in the listing of machine tools used by the Americans.[3] It would seem that the list of machine names in American plants was much longer than the corresponding list in British plants. Individual American machines were specialized – they could be used only for a small number of operations in a production process – but what they

[1] John D. Goodman, 'The Birmingham Small Gun Trade', in Samuel Timmins (ed.), *Birmingham and the Midland Hardware District* (London, 1866), pp. 392–3.

[2] Op. cit., p. 14. The figures on handicraft output, which represent an American estimate for the 1820s, are close to the British figures for the early 1850s cited by James Gunner, Superintendent of Enfield before its expansion. See *Report from the Select Committee on Small Arms*, op. cit., Question 4911.

[3] Nathan Rosenberg, 'Technological Change in the Machine Tool Industry, 1840–1910', *Journal of Economic History*, December 1963, footnote 18, gives some relevant citations of contemporary sources.

could do, they did much faster than the more versatile machines used by the British.[1] If contemporary observers can be believed, the distinctive feature of American machinery lay in its 'adaptation of special apparatus to a single operation'.[2]

Historians tend to insist upon the importance of individual kinds of machines in the characterization of a production process. Theorists tend to formulate a production process in such a way that the individual machines dissolve into a homogeneous 'jelly' (the term is Samuelson's) called 'capital'. The justification for either procedure lies in the degree to which it illuminates the subject under discussion. We argue that in an important class of cases which includes the Enfield Arsenal the specification of individual machines used is essential to an analytical treatment of the problem.

In the assertion, $y \leq f(x_1, \ldots x_n)$, made by a production function, the symbol x_i stands for a rate of utilization of the input whose name is i, for any $i = 1 \ldots n$. If two different machines may be associated with the same name i the same level of output may be obtained from every combination of machine-hours worked by these machines, providing only that the sum of the hours worked by the two is constant. When the British Parliamentary Commission comments on the length of the American list of machines we must ask whether more is involved than the mere names of machines.

Do the British use many machines with one name to achieve the same output that Americans achieve using one each of machines with different names? The Parliamentary Commission says that American machines (though less versatile than British machines) perform particular operations faster. That is, differing names are associated with differing performances. If we denote inputs common to British and American processes by the symbol $\bar{Z}$, then the American process, $y_A \leq f_A (\bar{Z}, x_{m+1},$

---

[1] E. Ames and N. Rosenberg, 'The Progressive Division and Specialization of Industries', *Journal of Development Studies*, July 1965, discusses the concepts of specialization and skill of inputs.

[2] 'As regards the class of machinery usually employed by engineers and machine makers, they are upon the whole behind those of England, but in the adaptation of special apparatus to a single operation in almost all branches of industry, the Americans display an amount of ingenuity, combined with undaunted energy, which as a nation we would do well to imitate, if we mean to hold our present position in the great market of the world.' *Report of the Committee on the Machinery of the U.S.A.*, op. cit., p. 32.

... x_n) is *not* the same as the British process $y_B \leq f_B (Z, x_{n+1})$. In this formulation the machine designated $(n + 1)$ in the British process replaces a collection of American machines, designated $(m + 1)$, ... n. If $\bar{X}$ is the total number of hours worked by some collection of American machines, suitably weighted,

$$\sum_{m+1}^{n} a_i x_i = \bar{X}$$

then the American machines are more productive than the English machines if $f_A(Z, x_1, \ldots x_n) > f_B (Z, \bar{X})$.

The adoption of machine processes, however, is usually more complicated than this example. 'The Blanchard lathe', which the Parliamentary Commission examined at the Springfield Arsenal, was really a sequence of sixteen different machines,[1] each of which spent only a very short period processing a given stock. These machines each replaced standard carpenters' tools and also the labour of skilled carpenters. But since the sequence of operations performed by Blanchard lathes does not correspond to the sequence of manual operations, we cannot associate particular lathes with particular carpenters or even particular carpenter tools. We may observe increases in the number of kinds of machines, decreases in the number of kinds of hand tools, increases in the number of machine operators and decreases in the number of carpenters required to produce a given number of guns, all in consequence of the adoption of 'the' Blanchard lathe.

A direct examination of the production process in an industry over time thus gives first-hand knowledge of technical changes in the sense of listing the inputs used; clearly in a micro-economic sense this examination enables the historian to characterize the difference in techniques used in (say) British and American gunmaking. Of course, this sort of comparison will not seem very satisfactory from a macro-economic point of view, for there is now no good way of analysing overall changes in the kinds of machinery and labour used.

Moreover, this information does not tell us whether differences in the list of machines used in two plants reflect differences in the prices of the machines, or whether they reflect

---

[1] *Report on the Committee on the Machinery of the U.S.A.*, op. cit., p. 39.

differences in the knowledge available to the management of the two firms.

Temin has ingeniously proposed the following problem in the theory of the firm: Suppose firms have production functions, which we cannot specify, because we do not have enough information. We then observe, over a period of time, the prices paid for inputs, the total cost of the inputs, and the rates of input utilization. Can we infer how the production function has altered over the period? If so, we have characterized 'technical change', in the firm or firms under study. The term 'ingeniously' appears in this paragraph because the proposal would involve using a body of literature in the theory of consumer demand. There the question is asked, 'Is it possible that a consumer who made some observed collection of purchases at various prices over a period of time can have had a constant set of tastes (utility function) throughout?' The problem of maximizing output for any given level of expenditures on inputs is formally the same as the problem of maximizing welfare for any given level of expenditures on goods, and it therefore constitutes a natural point of departure in cases where the production function is 'known' only in the general sense that utility functions are known.

Suppose we observe a consumer at two dates, and note the prices he paid, the goods he bought, and his total spending. Before asking anything about whether his tastes have changed, we imagine asking him, 'Were you better off at time 2 than at time 1?' If the axioms of consumer theory are valid the consumer can always answer this question by either a 'Yes' or a 'No' or a 'No difference'.[1] Analogously, if we look at a firm at two different dates, noting input prices, total outlays and input purchases, we must imagine that the firm can always answer the question, 'Did you produce more at time 2 or time 1?' Imagine that the firm can only answer, 'At time 1 I was making grommets, and at time 2 I was making hasps.' Then the question is meaningless.

The testimony of the Parliamentary Commissions indicates quite conclusively that Temin's question cannot be asked as a

[1] Consequently, the theorist would say that this problem does not admit of an answer if the consumer should reply, 'Well, the two situations are not comparable, from my point of view.'

means of comparing the Birmingham and Enfield processes, because the adoption of the Enfield process led a change in the nature of the product which was generally acknowledged to be of great importance. English buyers were either civilian or military. English civilians had their guns made to order, like their clothing, and were unwilling to purchase ready-made guns.[1] The English military, in 1811, had been left with 200,000 unusable muskets, because they did not have a system of inter-changeable parts.[2] By the time the Small Arms Committee met in 1854 even the most obtuse military mind (Lord Raglan, as Master-general, was the person charged with the responsibility for the supply of small arms) was convinced that rifles with interchangeable parts would enormously simplify the logistic problems of an army in the field. Thus all buyers would agree that the two alternative processes would produce different goods. A firearm possessing the property of interchangeability of parts was a very different commodity from one whose parts were not interchangeable. On this basis alone, Temin's pro-posed test could not meaningfully be carried out.

Economic theory is sometimes urged on historians because it allegedly is more precise. To confuse a gun with interchange-able parts with a handicraft gun, however, is to conduct analysis with less precision than any nineteenth century gunsmith, ordnance officer or Member of Parliament would have toler-ated.

English handicraft workers could not produce guns with interchangeable parts. Many English witnesses before the Small Arms Committee in 1854 maintained that interchangeability among gun components could not be achieved by any process at all. In particular, reports that highly irregular shapes, such as that of the gunstock, could be produced by machinery were met with general incredulousness.[3] Conversely, American

[1] 'The length, bend and casting of a stock must, of course, be fitted to the shooter, who should have his measure for them as carefully entered in a gun-maker's books, as that for a suit of clothes on those of his tailor.' Colonel Hawkins, as quoted in *A Treatise on the Progressive Improvement and Present state of the Manufactures in Metal* (London 1833), Vol. II, p. 105.

[2] Joseph W. Roe, 'Interchangeable Manufacture', *Transactions of the Newcomen Society*, XVII, p. 165.

[3] In his testimony before the Small Arms Committee Richard Prosser, an eminent Birmingham engineer, exhibited the handle of an American axe which had been made with the use of the Blanchard lathe. He told the Committee that

machine processes could not produce guns of the kind favoured by English civilians.[1] The Blanchard lathe produced stocks of a standard size, whereas English buyers did not want standard gunstocks. The English methods were suited to catering to the idiosyncratic needs of individual users. The Blanchard method was not suitable to filling individual orders.[2]

There is reason to believe that changes in the list of inputs used in manufacturing necessarily mean changes in the nature of the product. Nineteenth-century English observers frequently noted that American products were designed to accommodate the needs of the machine rather than the user. Lloyd, for example, noted of the American cutlery trade that 'where mechanical devices cannot be adjusted to the production of the traditional product, the product must be modified to the demands of the machine. Hence, the standard American table-knife is a rigid, metal shape, handle and blade forged in one piece, the whole being finished by electroplating – an implement eminently suited to factory production.'[3] In similar contrast, American guns were mainly used by farmers for utilitarian purposes.

To explain why America tended to use mass-production methods and England tended to use handicraft methods, one would have to make some hypothesis about demand in the two countries. Habakkuk[4] has spoken of a tendency for

---

although he had had the handle in his possession 'for many years' he could not persuade *anyone* that it had been machine made. *Small Arms Committee*, op. cit., Q. 2655. In fact, the Blanchard lathe had been used in the United States for many years before 1854 to produce irregularly shaped objects such as handles, hat blocks, oars, shoe lasts, spokes of wheels, etc.

[1] We would, of course, qualify 'could not produce' by the phrase 'at any reasonable price' if purists insisted.

[2] It was frequently observed that American machine techniques produced a less highly 'finished' product than the more labour-intensive methods. Americans, it was often noted, were more reluctant than the British to incur additional costs in ways which did not improve the efficiency of the final product in some narrowly defined utilitarian sense. The *Report of the Committee on the Machinery of the U.S.A.* pointed out this important taste difference with respect to American firearms methods. 'Many of the parts, after they are case-hardened, are polished on buffs, in the same manner as practiced in England; but on the whole much less attention is bestowed on what is called high finish given to the parts, only to please the eye; but that can be done, if necessary, after the parts are brought to shape by the machines' (op. cit., p. 38).

[3] G. I. H. Lloyd, *The Cutlery Trades* (London, 1913), pp. 394–5.

[4] Habakkuk, op. cit., p. 123. See also Rothbarth, op. cit., p. 386.

Englishmen to prefer variety and Americans quantity of goods. This statement remains intuitive. If it could be formulated precisely enough to admit of a test it would be of considerable interest to historians and theorists alike. The gunmaking industry is only one of a number of illustrations which could be brought to bear on the hypothesis.

One might, of course, continue to insist that the difference in 'quality' of guns was insignificant, and the structure of input costs the dominant factor in the Enfield decision. But within the next fifteen or twenty years Russia, Prussia, Spain, Turkey, Sweden, Denmark, and Egypt were all supplied with similar American gunmaking machinery.[1] The authors suspect strongly that the wage and interest rates in these countries exhibited very considerable variations, and consider this fact as strong evidence that machines made a different kind of gun from that made by handicraft workers.[2] This difference in product explains in large part the adoption of the new methods.

If technical change consists in the introduction of new kinds of machines and simultaneously of new kinds of products, then Temin's proposal for measuring technical change becomes unfeasible. For he wishes to consider a homogeneous output, made by homogeneous labour and homogeneous capital. Assume that indices can be constructed to cover the cases where output, labour, and capital are heterogeneous. These indices would be valid measures if the list of items which they include are fixed. But then they become invalid precisely at the points when technical changes occur.

The question remains: was the establishment of the Enfield Arsenal a response to changes in market conditions in the 1850s, or was it the result of a transfer of knowledge from America to England? In the first case an increase in the demand for mass-produced guns relative to handicraft guns meant that the optimal amounts of certain inputs (e.g. Blanchard lathes) rose, for the first time, above zero. In the second case, the production

[1] Fitch, *Report*, op. cit., p. 4. Japan and various Latin American countries bought finished American guns, but this fact is not relevant to arguments about the structure of input prices.

[2] If these countries had been free-traders we could use the factor-price equalization theorem to explain the decision. It is well known, however, that most of these countries had protectionist policies.

function for English guns was altered as a result of testimony before the Parliamentary Commission.

We have argued, so far, that genuine changes in English gunmaking production functions took place. Let us assume the role of Devil's Advocate. We can obtain, from the Parliamentary Commissions, some assertions that the British knew about machine methods of gunmaking but chose not to use them. If we accepted such assertions we could develop an argument consistent with micro-economic theory. This argument would rely upon considerations of uncertainty and instability of market demand. In applying it we who recognize the heterogeneity of 'guns' can use the evidence of the Commissions. But Temin, who aggregates output, cannot exploit evidence concerning product differences in explaining differences in gunmaking between the two countries.

For example, Richard Prosser (cf. footnote 3, p. 110) insisted in testimony to the Small Arms Committee that he would be willing to install gunmaking machinery if there were any evidence of stable rates of military procurement.

Q. 2651. If the introduction of machinery would facilitate the making of guns, why should not private gunmakers have introduced it? – There is not the slightest inducement for them to spend a penny in machinery; I would not do it.

Q. 2652. Why not? – Because there is no certainty that they would have government work to make that machinery remunerative.

Q. 2653. If a manufacturer had a contract for a long time, do you think that machinery could be introduced with advantage by private gunmakers? – No question about it; if you gave them 500,000 to make a year, there would be no difficulty in finding money to erect machinery.

Q. 2654. If you had that number of guns to make a year, would you introduce machinery? – I could not make them without.[1]

[1] *Report from the Select Committee on Small Arms*, op. cit. Prosser also cited the example of one of his employees, to whom he paid a wage of 35*s.* a week, turning down a weekly wage of £3 in the gun trade because of the uncertainty of employment prospects in that trade (Q. 2654 and Q. 2655). Similar statements may be found throughout the testimony. Cf. Q. 5415–Q.5417. The Committee in its final report, drew conclusions similar to those of Prosser: 'There is . . . reason

In the absence of steady orders, British gunmakers preferred (rationally) to pass the risk of low orders on to their employees rather than incur it themselves. This preference may explain the predominance, in England, of the (labour-intensive) handicraft system.[1] The British Government apparently did little, if anything, to moderate the extreme instability which its contracting procedures imposed upon the gunmaking industry, or to enable the industry to operate with a longer-term planning horizon. The United States Government, in contrast, gave out long-term contracts which enabled contractors to maintain steady rates of output, and hence to buy machines without undue risk. Indeed, in the early years the financial advances given by the United States Government to contractors constituted a major source of capital to contracting firms, many of which (such as that of Whitney)[2] became important innovators in the new machine technology.[3] Perhaps of even greater importance was the nature of civilian markets. Standardization of parts may have met with less resistance in America, because American farmers needed large numbers of guns for utilitarian purposes. In England, where natural predators were fewer, firearms were mainly an upper-class

---

to believe, that if the gun-trade could have confidence that a large supply of muskets would be purchased by the Ordnance in future years, the manufacturers themselves would be anxious to introduce machinery wherever it could be profitably employed.' Ibid., p. vii.

[1] 'Q. 3539. Do you know that very frequently men who have been employed at the Government work, have had to stand still 15 or 18 weeks together without work, in consequence of the irregularity of the demand? – I frequently hear complaints from the men that sometimes they have nothing to do and are starving, because the Government contracts are so dilatory.' Ibid. Testimony of John Dent Goodman, Birmingham merchant and partner in a gun manufactory.

The interested reader is referred to Stigler's article on optimization in the short run which examines, within a theoretical framework, a range of problems similar to those confronting British gunmakers. George Stigler, 'Production and Distribution in the Short Run', *Journal of Political Economy*, June 1939, pp. 305–27.

[2] J. R. T. Hughes, *The Vital Few* (Boston: Houghton-Mifflin, 1966), Chapter 4, discusses these matters at some length.

[3] Felicia Deyrup concludes, in her careful study of the New England firearms industry: 'the contract system was of immense value both to the government and the contractor, for aside from bringing the industry into existence it promoted a spirit of cooperation and mutual aid unique among early American manufacturers, which had much to do with the rapid development of the industry in the first thirty years of the nineteenth century.' Felicia Deyrup, *Arms Makers in the Connecticut Valley*, Smith College Studies in History, Vol. XXXIII (Northampton, Mass., 1948), p. 66.

commodity, and total civilian demand was therefore relatively small.

Moreover, the Enfield Arsenal was equipped with a new and improved line of stocking machinery which had been designed by Cyrus Buckland and introduced by the Ames Manufacturing Company of Chicopee, Mass., in 1853.[1] It might therefore be the case that the earlier models of American stocking machines, at least, would have been unprofitable under British conditions prior to 1853. To accept this view, however, means disregarding a great deal of evidence, throughout the hearings and elsewhere, that English gunmakers simply did not consider it possible to make guns with interchangeable parts, and did not believe that the Americans had been making such guns by machine methods.

Suppose that the Parliamentary Commission changed English gunmaking techniques as effectively (and in much the same way) as Eli Whitney, Simeon North, and John Hall had earlier changed United States gunmaking. In this case, the historian will look to the theorist for a hypothesis about how new inputs 'enter' a production function, and how obsolete inputs 'go out of use'. Unfortunately, he will not find much help at present. The production functions most commonly used do not readily 'admit new inputs'; and theoretical analysis regards zero utilization of any inputs as an inconvenience.[2] The whole thrust of modern theory is quantitative, exploring the relative amounts of inputs and outputs. Qualitative information on the use or non-use of inputs fits awkwardly into such theory; in contrast,

[1] Fitch, *Report*, op. cit., p. 15. Temin is therefore wrong when he states, in reference to the British purchase of American stocking machinery: 'The American machines did not represent the latest technological developments, and British enthusiasm for them remains a mystery.' Temin, 'Labor Scarcity and the Problem of American Industrial Efficiency in the 1850s', op. cit., p. 282. In the previous sentence he states that the Blanchard machine 'had been used "very extensively" for thirty years prior to the English visit'. Although there are some assertions that Blanchard-type stocking lathes had been tried out experimentally on some occasions, we are prepared to state categorically that there is no evidence supporting the proposition that they were used 'very extensively' in England in the thirty years prior to 1854. See *Small Arms Committee*, op. cit., Questions 2655–8, 4191–2, 4291–4, and 4313–20.

[2] Theory explains some 'equilibrium conditions' neatly. A price exactly high enough to induce zero utilization of a given input is an equilibrium price; a price higher than that will also induce zero utilization without being, in the same sense, an equilibrium price. Where some equilibrium conditions are equalities and others are inequalities, theory becomes messier.

economic history makes much use of factual information which is not quantitative.

The historical problem of explaining the impact of the Parliamentary Commissions on gunmaking is not unique, nor even contrived. Consider the analogous problem of accounting for changes in gunmaking over time. As of 1785, neither the British nor the Americans could make guns with interchangeable parts. As of 1815, Americans could make guns with interchangeable metal parts, but could not make interchangeable gunstocks.[1] As of 1820, they could make interchangeable gunstocks. At any date, presumably, they could use not only current methods but earlier methods which these had displaced.[2] There is no economic theory at present which represents a production function which copes explicitly with this type of phenomenon, even though it is very often observed in history.[3]

Whether we conclude that the English of 1850 could have made guns in the American way, but chose not to, or whether we conclude that English production functions changed as a

[1] The earliest attainments of interchangeability did not involve extensive use of machinery. Rather, uniformity was assured by the use of jigs and fixtures. Of course, the degree of uniformity attainable by such methods as hand filing with hardened jigs did not approach what was later attainable, say, with a milling machine. As Fitch observed, comparing the uniformity of 1880 with standards of uniformity early in the century: 'If parts were then called uniform, it must be recollected that the present generation stands upon a plane of mechanical intelligence so much higher, and with facilities for observation so much more extensive than existed in those times, that the very language of expression is changed. Uniformity in gun-work was then, as now, a comparative term; but then it meant within a thirty-second of an inch or more, where now it means within half a thousandth of an inch. Then interchangeability may have signified a great deal of filing and fitting, and an uneven joint when fitted, where now it signified slipping in a piece, turning a screwdriver, and having a close, even fit' (Fitch, *Report*, op. cit., p. 2).

[2] Producers have finite memories. Gunmakers of 1967 would find it inordinately difficult to make guns by the methods of 1785. However, 'Roman' chariots used in motion pictures of 1967 are probably superior, by Roman standards, to first-century chariots.

[3] Space does not permit a full documentation of this point. The reader is referred to Herbert Simon, 'Effects of Technological Change in a Linear Model', in T. C. Koopmans (ed.), *Activity Analysis of Production and Allocation* (New York, 1951). Here some inputs and outputs are said to be used in zero quantity, and some to have price zero at any moment. The list of such inputs and outputs changes in response to the set of 'activities' used. In various important ways, however, it appears to us from Simon's discussion (p. 262) that the list of inputs and outputs remains unchanged as 'technology' changes. This important paper is difficult to use as a basis for empirical analysis, and we feel that theory has yet to come to grips with the sort of problem represented by our example.

consequence of the Parliamentary Commissions, we deal with phenomena which are not treated adequately by current economic theory. Since economics ultimately deals with real things, it is natural that historians should present new and unsolved problems for theorists. If, then, they propose the problem, 'Why did American gunmaking develop interchangeability of parts and machine production methods long before British gunmaking?' they are merely providing the theorist with subject matter for future study. In this sense, they are the substitute for experimental scientists in a discipline which does not generally admit of experimentation.

It is a commonplace among theorists that the use of aggregated models involving inputs like 'capital' and 'labour' are mainly useful as exercises to tell theorists something of the intuitive nature of theories whose properties, when fully elaborated, may not be obvious. When such hypotheses are used in empirical work great care must be used to explain the procedure used in combining the inputs (and, of course, the output if several are involved). The burden of proof is upon the user of these theories, since he must explain why it is reasonable to proceed in this fashion.[1]

Economic historians have been reluctant to use economic theory, partly because of their preoccupation with particular events, and partly because of what is evidently a mistrust of abstraction. In the present problem it appears that the characteristics of individual machines and their outputs are essential parts of the subject matter under discussion. If this is so, then the use of drastically aggregated input hypotheses is apt to be

[1] P. A. Samuelson, 'A New Theorem on Nonsubstitution', in Hugo Hegeland (ed.), *Money, Growth and Methodology* (Lund Social Science Studies, No. 20, Lund, 1961), states: 'Economists often . . . try to talk of primary factors like labor and land, *and* some *other* aggregative factor called "capital". All this, it bears repeating, is but a figure of speech. . . . In *perfect* competitive markets . . . no one can stop the contented merchant from adding up his asset values, or prevent the national statistician from adding up his estimates of the community's wealth. Such computed value totals, it need hardly be said, cannot be fed into production functions, kicked or leaned against' (p. 410). 'The modern theorist has no real need for such childish devices [as an 'abstract capital'] and, in any case, should not be permitted to use them until he has shown he can rigorously formulate his theories completely without them' (p. 413). 'Reality . . . involves great differences in natural resources between countries, great differences in technology or effective technology, and great differences in achieved capital formation' (p. 414). Italics are Samuelson's.

misleading. In this kind of case at least, the mistrust of the historian may be justified.

Properly stated, economic theory is obviously useful to economic historians. No one can work without hypotheses, and many parts of economic theory are well suited to the analysis of historical problems. There are cases, however, where the existing theoretical literature is not suitable for the discussion of historical problems. The theory of technical change is one case in point. As it has developed, it relies upon highly aggregated input and output data.

One can tell that this theory is not suited to historical material in the following way. When one tries to use the historical record – in this case the Parliamentary records – one discovers that the evidence presented does not have a place in the theoretical structure. It is not possible to talk about Blanchard lathes and carpenters' tools in the context of homogeneous capital; about machinists and craftsmen in the context of homogeneous labour; about sporting guns and guns with interchangeable parts in the context of homogeneous output. Indeed, one cannot even state the proposition, 'The English did not know how to mass-produce guns', in the context of a production function with inputs of homogeneous labour and homogeneous capital.

The central issue in the historical literature on technical change in the nineteenth century seems to be this: Americans clearly led the British in the adoption of many machine methods of production. If this precedence is not simply 'Yankee ingenuity' working in a void it must reflect such economic factors as resource endowment, the structure of the labour force, the structure of prices and the nature of consumer tastes. The simplest techniques of analysis reveal that several variables must be considered simultaneously. The working historian will naturally wish to keep his explanation as simple as he can. Even when he works on a very narrow problem, such as the establishment of the Enfield Arsenal, however, he must perforce involve himself in a variety of questions: the nature of the demand for the product, the nature of the change in the list of inputs, the way in which the various inputs affect costs and so on. Fundamentally, he may wish to keep his list of variables short, but on historical and analytical grounds he knows that every time he suppresses a variable a part of his record becomes irrelevant.

We do not know whether American gunmaking differed from English gunmaking because of differences in consumer tastes, because of differences in knowledge about machine design, because of differences in the training of the labour force, or because of differences in the structures of factor prices. It is legitimate to try to assess the relative importance of these differences. Such an assessment requires the utilization of historical evidence. This evidence is in many cases qualitative, not quantitative; an event occurred, or an event failed to occur. The emphasis in theoretical work has been mainly quantitative, and in cases where the historical record is quantitative, considerable progress has been made in unifying economic history with economics generally.

The record, however, is not always so obliging. Qualitative changes – the appearance of new products and processes, for example – must be treated by means of qualitative theories. It will not do for theorists to brush aside such intractable aspects of the historical record, for these qualitative changes may well constitute the very essence of the phenomena. If so, historians are to be praised for forcing them upon the attention of reluctant economists.

4. American Rings and English Mules: The Role of Economic Rationality[1]

LARS G. SANDBERG

[This article was first published in the *Quarterly Journal of Economics* (February 1969).]

The single most important technological improvement in the cotton spinning industry during the last hundred years has been the replacement of mule by ring spinning. Most of the technical development of this process took place in the United States, and it was in the United States that it first rose to economic prominence.

As early as 1870 ring spinning had become the dominant form of spinning in the United States. In that year there were a total of 3·7 million ring spindles and 3·4 million mule spindles installed in the United States.[2] Since, for a given fineness (count)[3] of yarn, output per ring spindle exceeds output per mule spindle, a comparison of the numbers of the two types of spindles installed tends to understate the importance of ring spinning. By 1905, 17·9 million of the 23·1 million installed spindles in the United States were rings.[4] This trend, of course, continued and by the outbreak of the Second World War mule spinning was virtually extinct in the United States.

Ring spinning was also introduced into other parts of the world, but at a slower pace than in the United States. If modern

[1] I am indebted to Alexander Gerschenkron and the members of the Economic History Seminar at Harvard for suggestions and encouragement. A somewhat longer and more detailed version of this paper is available from the author on request.

[2] Melvin T. Copeland, *The Cotton Manufacturing Industry of the United States* (Cambridge, Mass.: Harvard University Press, 1912), p. 70. For a technical description of mule and ring spinning, see John Jewkes and E. M. Gray, *Wages and Labour in the Lancashire Cotton Spinning Industry* (Manchester, England: Manchester University Press, 1935). Chap. 1.

[3] The 'count' of a yarn is defined as the number of hanks, at 840 yards each, per pound.

[4] Copeland, op. cit., p. 70.

cotton industries outside the United States are divided into three general groups, non-European (principally those in Japan, China, and India), Continental European and British, then the percentage of ring spindles installed at any given point in time decreased in the order listed above. That is, Great Britain, with the largest cotton industry in the world, was last among all important cotton industries in the introduction of ring spindles.[1] As late as 1913 there were 45·2 million mule spindles but only 10·4 million ring spindles in Great Britain.[2]

Although some contemporary observers noted that Great Britain had certain special advantages in mule as opposed to ring spinning,[3] the British lag in ring spinning has usually been taken as a sign of technological conservatism, not to say backwardness.[4] This view has been reinforced by subsequent developments. Ring spinning has indeed proven to be the wave of the future and the British cotton textile industry has experienced a sharp decline ever since the end of the First World War. Neither of these later trends, however, in any way proves that the British made a mistake, or were irrational, in not introducing more ring spindles before the First World War. Under the conditions then prevailing with regard to factor costs, as well as the technical capabilities of the ring spindles then being built, the British may well have been acting rationally.

The question whether the difference in the ratio of rings to mules in Great Britain and the United States in some given year was justified by differing factor costs and market conditions is, however, extremely difficult to answer. Phrased in this form, the question presents several formidable obstacles to quantitative analysis. First of all, as will be discussed below, the relative efficiency of ring and mule spindles varied for different counts of yarn. Thus, a detailed knowledge of the counts of yarn spun in the United States and in Great Britain would be needed. Although it is generally presumed that Great Britain devoted a larger percentage of her spindles to high quality

[1] R. Robson, *The Cotton Industry in Britain* (London: Macmillan, 1957), p. 355.
[2] Ibid.
[3] Copeland, op. cit., pp. 71-3.
[4] See, for example, Rockwood Chin, *Management, Industry and Trade in Cotton Textiles* (New Haven: College and University Press, 1965), p. 85.

yarn than did the United States,[1] no sufficiently detailed information is available for the pre-First World War period. Furthermore, installing rings when a new plant was built or when old mules were physically worn out, was quite a different thing from throwing out technically well-functioning mules and replacing them with rings. Thus, the optimal mix of rings and mules depended not only on the distribution of counts spun, but also on the past rate of expansion of the industry.

The situation is further complicated by differences in factor costs in the two countries. These differences mean that the profitability of replacing mules by rings on a given count differed in the two countries. Not only does this mean that the same count might sometimes rationally have been spun by different methods in the two countries, but also that mules would be judged ready for the scrap heap at different ages in the two countries. There is thus no doubt that in the pre-First World War period it paid to keep old mules longer in Great Britain than in the United States. Any calculation of this effect on the optimal combination of rings and mules in the two countries would require a detailed knowledge of the distribution of mule spindles by age in the two countries, as well as a huge amount of information about the effect of age and obsolescence on the costs of mule spinning. In addition, of course, the unavailable information on the distribution of counts spun would be needed. Clearly, these obstacles require that the original question be rephrased in a more convenient form.

Since it is clear that rings were considered more suitable for low- as opposed to high-count yarn, I have decided to concentrate my attention on the counts at which new investment generally shifted from mules to rings in each of the two countries. That is, the central question of this paper will be, can rational economic forces explain why American firms generally installed rings to spin all yarns up to count X while British firms continued to install mules for counts lower than X?

The date for which this question will be investigated is the period immediately preceding the First World War. This period has the advantage that a good deal of information is available for it. Unfortunately, some of the information on the technical characteristics of mules and rings needed for this study is

[1] See, for example, Copeland, op. cit., p. 71.

available only for a later period. This later period is not a good one for the study as a whole, however, principally because the extremely depressed conditions that have existed since the First World War are clearly not conducive to a study of investment behaviour and technological change.

OBSERVED INVESTMENT BEHAVIOUR

The first problem of this study is to determine whether there really were fairly sharp cutoff points between rings and mules when new spindles were being installed in the United States and Great Britain, respectively, and if so, at what counts these cutoff points occurred. In his excellent study of the American cotton textile industry published in 1912, Melvin Copeland reports that 'Not much yarn finer than 40's and very little higher than 60's is produced upon the ring-frame in Europe, whereas practically all warp yarn, even up to 120's, is spun upon that machine in America.'[1]

This statement is well supported by other evidence available from the same period. The works by Uttley and Young are full of examples of high-quality yarn spun on rings in the United States.[2] Not only is this true with regard to warp, it also seems to be the case for weft,[3] although the references to high-quality weft being spun on mules are somewhat more common than those to high-quality warp yarn. On the other hand, there are no references to any yarn below 40 being spun on mules.[4]

As for new installations, the U.S. Department of Commerce reports that between 1900 and 1914, only 981,023 new mule spindles were installed in the United States as opposed to 11,888,587 new ring spindles.[5] Further confirmation of the

[1] Copeland, op. cit., p. 301.

[2] T. W. Uttley, *Cotton Spinning and Manufacturing in the United States of America* (Manchester, England: Manchester University Press, 1905), pp. 9, 11, 16, 22, 23, 29, 31, 32, 34, 49, 54, 56, and 60, and T. M. Young, *The American Cotton Industry* (London: Methuen, 1902), pp. 10, 16, 18, 19, 24, 35, 61, 68, 73, 86, 88, 97, and 110.

[3] Warp, also known as twist, is the yarn which is stretched in the loom. Weft, also known as filling, is the yarn inserted into the warp by means of the shuttle. Warp has to be stronger than weft.

[4] See previous references in Uttley and Young.

[5] U.S. Department of Commerce, Bureau of Foreign and Domestic Commerce, *The Cotton Spinning Machinery Industry*, Miscellaneous Series, No. 37 (Washington, 1916), Table 44, p. 77.

low number of new mule spindles being installed can be obtained from the U.S. Census of Manufactures for 1914 where it is stated that 'the installation of these (mule) spindles has practically ceased'.[1] It is clear that the small number of mules still being put in were intended to make very high-quality yarns. In fact, the *Census of Manufactures, 1905*, concluded that the only reason any mules at all were being installed was that 'there are some (high) qualities of yarn which cannot be made successfully by ring spinning'.[2]

The most important evidence that rings were being installed in Great Britain for yarns up to the lower 40's but very seldom above that range comes from the Universal Wage List for Ring Spinning which was adopted in 1912 and which covered virtually all British ring spinning.[3] Because the number of spindles tended by a spinner increased as the count of the yarn being spun increased, the list was designed to give a lower piece rate per spindle as the count spun increased. This accommodation is made for counts up to and including 43, but then stops abruptly. This is true despite the fact that the tendency towards more and more spindles per spinner continued on past the 40's and that the adjustment would have created no great computational problems.[4] The only reasonable conclusion is that the list ends because there were virtually no spinners working on higher counts.

The evidence thus makes it clear that at least some rings were being installed for counts up to the lower 40's, but virtually none for counts above that. This is of little value, however, unless it can at least be shown that installations below 40 were not unusual occurrences. Fortunately such evidence is available.

Between 1907 and 1913, the number of installed mule spindles in Great Britain increased from 43·7 million to 45·2 million, and the number of ring spindles increased from approximately 8·3 million to 10·4 million.[5] On the basis of several reasonable assumptions, it can be shown that these new rings were

[1] U.S. Bureau of the Census, *Census of Manufactures, 1914*, II, 38.

[2] Ibid., *1905*, III, 42.

[3] Jewkes and Gray, op. cit., pp. 117 and 128.

[4] Ibid., p. 121.

[5] British Census Office, *Census of Production, 1907* (London: His Majesty's Stationery Office, 1908), I, 293 and Robson, op. cit., p. 355.

enough to account for all the increase in sub 40 capacity and to replace about 15 per cent of all the mules used for sub 40 yarn in 1907.[1] At the very least, it is clear that a very large percentage of the spindles installed for counts up to 40 were rings.

THE RELATIVE ADVANTAGES OF RINGS OR MULES IN THE UNITED STATES AND GREAT BRITAIN

The purpose of this section is to examine the differences that existed in the benefits to be derived from replacing mules with rings in the United States and the benefits to be derived from doing so in Great Britain. The discussion will be divided into two parts, the first dealing with factor costs and the second with such problems as the role of labour unions and the technological interrelationships between ring spinning and automatic weaving.

A. *Factor Costs*

(1) *Labour Costs* The principal advantage of ring as opposed to mule spinning was that the former used unskilled or semi-skilled female labour while the latter used highly skilled males. Good estimates of the spinning labour costs of the two methods of producing yarn in this period are available for both the United States and Great Britain in Copeland's book which presents a range of spindles per operator and a range of wages per operator.[2] I have focused attention on the high estimates for both the number of spindles tended and the weekly wage. In fact, it would not matter very much if I had used the lower estimates for both, or if I had chosen a middle position. I use the upper limits for both number of spindles and wages because these figures are most likely to apply to the new equipment in which I am principally interested.

Taking account of the difference in ring spindle speed in the United States and Great Britain and allowing for the difference in productivity between ring and mule spindles at a count around 40, the spinning labour cost for ring spinning turns

---

[1] For a detailed discussion of this problem, see the author's longer paper which is available on request.

[2] Copeland, op. cit., pp. 298–300.

out to have been about 50 cents per week, per hundred 'mule equivalent' spindles[1] in both countries. For mule spinning the cost was around $1.65 per hundred actual mule spindles per week in Great Britain and $2.15 in the United States. This, in turn, implies that at a count of 40, ring spinning labour per pound of yarn was about 1·6 cents lower in Great Britain and about 2·4 cents lower in the United States than the cost of mule spinning labour.

(2) *Capital* Mule and ring spinning appear to have been of almost exactly the same capital intensity per unit of output in the production of yarns of a count around 40.[2] Below this count, mules tended to be more capital-intensive and above it rings were more capital-intensive. This rough equality around 40 was the result of higher machinery costs for ring spinning, including some extra roving equipment, offset by the space saving achieved in ring spinning. In view of this fact, it is difficult to believe that any difference in interest rates could have had much to do with America's greater propensity to install mules. Certainly, if rings stopped being profitable in Great Britain at a count of 40, no reasonable change in the interest rate could have made them profitable.

It does seem apparent, however, that mule spinning machinery was more expensive relative to ring spinning machinery in the United States than in Great Britain. Evidence for this comes primarily from the fact that between 1900 and 1914, 77·6 per cent of all new mule spindles installed in the United States were imported from Great Britain, while only small quantities of other types of cotton textile machinery were imported.[3] This, of course, was not due to any inherent inefficiency in mule making in the United States, but rather to the fact that so few mules were being installed that it was not profitable for American producers to make mule spinning frames.[4]

---

[1] I have converted ring spindles into mule equivalence so that the cost comparisons can be based on equal quantities of output.

[2] James Winterbottom, *Cotton Spinning Calculations and Yarn Costs* (2nd ed.; London: Longmans, Green and Co., 1921), pp. 213, 272, and 273. Winterbottom was lecturer in cotton spinning at the Municipal School of Technology in Manchester during the period covered in this study.

[3] *The Cotton Spinning Machinery Industry*, Table 44, p. 77.

[4] Since America was a high-cost, protected producer of textile machinery, there was no hope of capturing an export market for mule frames.

While it is difficult to tell how much difference there was in the relative prices of the two types of machinery, this effect must have given some further impetus to ring spinning in the United States. On the other hand, it should be remembered that ring spinning tended to save capital in the form of buildings and required extra capital for machinery. In view of the fact that construction was relatively cheaper as compared with machinery in the United States than in Great Britain,[1] this would tend to favour the use of the 'construction-intensive' mule process in the United States.

(3) *Fuel and Lubricants* Here, again, the costs appear to be virtually the same for the two methods. Sources can be found that disagree about which method saved fuel.[2] In any case the difference was very small when expressed in terms of cents per pound of yarn. This means, of course, that any effect of different fuel prices in the two countries on the choice of spinning technique must have been infinitesimal.

(4) *Transportation* Transportation is treated as an input because the yarn had to be moved before it could be woven into cloth. The difference in transportation costs between rings and mules arises because mule yarn was spun either on a bare spindle or on a paper tube, while ring yarn had to be wound on a heavy wooden bobbin. Fortunately, the warp yarn could be rewound. The weft, however, had to be shipped on the bobbin.[3] Copeland quotes with approval an estimate that the paper tubes added only 10 per cent to the freight costs, while the wooden bobbins added 200 per cent. Furthermore, the bobbin had to be returned.[4]

The reason this difference in transportation costs affected Great Britain and the United States differently is that the American industry was vertically integrated while the British industry was not. Thus, much more yarn transportation was

[1] Young, op. cit., p. 9.
[2] Winterbottom, op. cit., pp. 272 and 273, and W. A. Graham Clark, *Cotton Textile Trade in the Turkish Empire, Greece, and Italy*, Bureau of Manufactures, Special Agents Report No. 18 (Washington: U.S. Government Printing Office, 1908), pp. 89 and 90.
[3] Copeland, op. cit., pp. 69 and 72.
[4] Ibid., p. 69.

required in Great Britain than in the United States. In addition, Britain had a large export trade in yarn.

Fortunately, a good estimate can be made of the level of these extra transportation costs. Reliable information is available on the cost of shipping yarn in Lancashire in 1907 over the average distance yarn was in fact shipped.[1] If the 200 per cent cost increase figure is used together with an allowance for the extra cost of returning the wooden bobbins, it appears that shipping ring weft within Lancashire cost about three mills more per pound of yarn than shipping mule weft. The cost differential, of course, did not apply to yarn produced in integrated plants.

It is impossible to give a single cost estimate for exports, since it depended on the destination of the yarn. This extra cost of exporting ring rather than mule weft, however, could clearly be very high or even prohibitive. Nevertheless, the effect of this cost differential is severely limited by the fact that in this period only 10 to 15 per cent of total British yarn production was being exported.[2] This means that no more than 5 to 8 per cent of all the yarn produced was weft for export. Since there were many more mules working at all counts than were needed to produce this percentage, there was probably little effect on investment behaviour. Presumably, the export transportation disadvantage of ring weft resulted mainly in the concentration of the production of sub 40 weft for export on mules installed in the pre-ring period.

B. Other Factors Affecting the Choice of Spinning Method

(1) *Labour Unions* The literature on the history of the American cotton textile industry is full of references to the effect that the disruptive and belligerent attitude of the American mule spinners' unions was a major factor in encouraging the shift to ring spinning. Thus, it is reported that it was after the strike of January 1898 (led by the mule spinners) that the treasurer of the highly efficient and well-managed Pepperell Manufacturing Company, of Biddeford, Maine, 'made plans to get rid

[1] William Whittam, *Report on England's Cotton Industry*, Bureau of Manufactures, Special Agents Report No. 15 (Washington: U.S. Government Printing Office, 1907), p. 32.
[2] See Robinson, op. cit., p. 333.

of all the mule frames eventually and to put in ring frames.'[1] Earlier, the cotton manufacturers of Fall River, Massachusetts, had been 'particularly anxious to introduce ring spindles' as a result of the strikes of 1870 and 1875.[2] A more general comment to the same effect appears in the *Census of Manufactures,* 1905:

> But there are reasons, not unconnected with the labor problem, which render manufacturers desirous of using frames (i.e., rings) rather than mules whenever it is (technically) practical to do so.[3]

These observations must, however, be viewed with at least some reservations. In all cases there were other good reasons for introducing rings. Furthermore, the manufacturers had every interest in making the workers fear that aggressive union action would result in technological unemployment. Nevertheless, there can be no doubt that American manufacturers had a strong aversion to unions and that the mule spinners were probably the most efficient and powerful cotton textile union, at least until the absolute number of installed mule spindles started to decline around 1900. It is also clear that the mule spinners' union tended to encourage at least temporary organization among the other workers and that it was largely responsible for many strikes.[4] Thus, the desire to break the power of the union by replacing the obstreperous mule spinners with docile girl spinners probably did have at least some effect in encouraging the adoption of ring spinning in the United States.

In the case of Great Britain, there was also a sharp contrast between the powerful and well-organized mule spinners and the weakly organized ring spinners. The British employers, however, appear to have been better adjusted to the fact of having to face unions than were American employers. The British mule spinners' union was far from the only strong British cotton

[1] Evelyn H. Knowlton, *Pepperell's Progress* (Cambridge, Mass.: Harvard University Press, 1948), p. 171.

[2] R. Smith, *The Cotton Textile Industry of Fall River, Massachusetts* (New York: King's Crown Press, 1944), p. 100. See also Robert K. Lamb, 'The Development of Entrepreneurship in Fall River, 1813–1859', unpublished Ph.D. thesis, Harvard University, p. XII–8.

[3] *Census of Manufactures, 1905,* III, 42.

[4] Knowlton, op. cit., pp. 170–1, and Smith, op. cit., p. 100.

union. Even more important, the British mule spinners were mainly dedicated simply to raising their own wages. To the extent they succeeded in this endeavour, they may, of course, have helped the cause of ring spinning, but any such effect has already been considered in the section on relative wages.

(2) *Relation of Ring Yarn to Automatic Looms* The period under discussion in this paper was also a period during which large numbers of automatic looms were installed in the United States. Automatic looms, however, or at least these automatic looms, required the greater strength of ring as opposed to mule yarn.[1] This complementarity between ring spinning and automatic weaving meant that the existence of ring spinning made the introduction of automatic looms more appealing and, inversely, plans to install automatic looms depended on the availability of ring spinning.[2] There may, therefore, have been some American manufacturers for whom a desire to introduce automatic looms made ring spinning relatively more advantageous as compared with mule spinning than would otherwise have been the case. For the most part, however, ring spinning clearly preceded automatic weaving.[3] With regard to Great Britain, this interdependence between ring spinning and automatic looms can be ignored for purposes of this paper since automatic looms did not begin to appear there in significant numbers until the 1930s.[4]

THE ROLE OF COTTON PRICES

This discussion of factor costs and other considerations has shown that in virtually every category the advantage of replacing mules by rings was greater in the United States than in Great Britain. This fact combined with the generally accepted fact, to be discussed in greater detail below, that the relative advantage of ring as opposed to mule spinning declined as the count spun increased, generally accords well with

[1] Irwin Feller, 'The Draper Loom in New England Textiles, 1894–1914: A Study of Diffusion of an Innovation', *Journal of Economic History*, XXVI (Sept. 1966), 331.

[2] Ibid., p. 333.

[3] See Copeland, op. cit., pp. 70 and 87, and Robson, op. cit., p. 355.

[4] In 1937, only 3 per cent of all British cotton looms were automatic. Robson, op. cit., p. 210.

the observed fact that Great Britain stopped installing rings at a count of about 40, while the United States continued installing rings at much higher counts. It says very little however, about whether the British cut-off line logically should

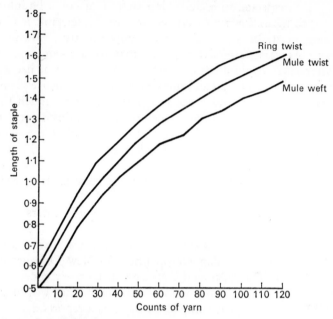

Source: Winterbottom, p. 235.

Note: Figure 1 is based on the following equations:

Length of staple in inches, for ring twist $= 0.35 \ (\sqrt[3]{\text{count}})$.

Length of staple in inches, for mule twist $= 0.325 \ (\sqrt[3]{\text{count}})$

Length of staple in inches, for mule weft $= 0.30 \ (\sqrt[3]{\text{count}})$

Figure 1

COTTON STAPLE LENGTHS 'SUITABLE' FOR THE SPINNING OF
VARIOUS YARNS

have been drawn exactly where it was drawn. Some information on this problem can be obtained from the structure of cotton prices.

Cotton prices played an important role in the choice of spinning technique because of the technological fact that, for a given count of yarn, ring spinning required a longer cotton staple than did mule spinning. Figure 1 is designed to show which lengths of fibre were 'suitable' for different counts.[1]

Unfortunately, the compiler of the information used to produce Figure 1 considered only ring twist, mule twist, and mule weft. He neglected to include ring weft. This could be interpreted to mean that ring weft required the same staple length as ring twist. This was almost certainly not true, however. Most observers were of the opinion that, for a given count, a shorter staple could be used to produce weft than was needed for twist, both on mules and rings.[2] Indeed, the evidence seems to indicate that the difference in length was pretty much the same regardless of the spinning method used.[3] If this is correct, it means that the mule twist requirements would be the same as the ring weft requirements. I will, therefore, treat the difference between the staple length needed for mule twist and mule weft as representing the difference between ring weft and mule weft.

This difference in the required staple length enters as a factor in the choice of spinning technique the price of cotton generally increased as its staple length did. This is shown in Figure 2 which contains cotton prices in New Orleans on 1 April 1913. This year and date were deliberately chosen as representing a season and period when the market was 'normal.'[4] In particular, the compiler of these data reports that the big jump in price occurring between staple lengths of $1\frac{1}{16}$ in and $1\frac{1}{8}$ in was 'common at all times.'[5] The exact size of the price jumps between different staple lengths must have varied somewhat depending on harvest conditions as well as peculiarities of final demand for cotton products, but Figure 2 can certainly be taken as representative of the period just preceding the First World War.[6]

Combining Figures 1 and 2, it is possible to compute the

[1] Winterbottom, op. cit., p. 236.

[2] Paul H. Nystrom, *Textiles* (New York: D. Appleton, 1916), pp. 71–2.

[3] Ibid.

[4] Winterbottom, op. cit., p. 234.

[5] Ibid.

[6] Figure 2 does not include the very longest staple cottons. It should thus be added that at the very top of the scale the jump was from cotton of about $1\frac{3}{4}$ inches to about 2 inches, with virtually nothing in between, and an increase in price amounting to around 5 or 6 cents per pound at July 1914 prices. John A. Todd, *The World's Cotton Crops* (London: A. & C. Black, 1915), p. 17.

differential cotton cost in spinning with rings as opposed to mules. A literal application of the technical information in Figure 1 results in Figure 3 for ring twist versus mule twist and Figure 4 for ring weft versus mule weft.

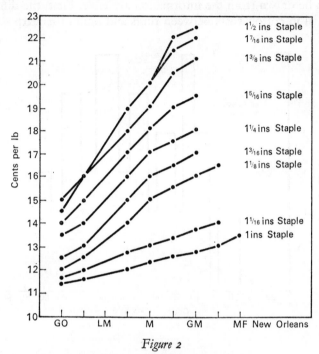

Figure 2

PRICES OF VARIOUS COTTONS BY QUALITY
AND STAPLE LENGTHS, NEW ORLEANS, 1 APRIL 1913

It is clear that Figure 1, and therefore Figures 3 and 4 are primarily based on technological rather than economic considerations. Since cotton prices were not quoted continuously by length, but by steps of $\frac{1}{16}$ in, rational producers would be prepared to accept somewhat higher costs in order to avoid the next step on the staple progression. Rather than immediately going to the longer staple when the count they were spinning required it if they were to continue with the exact production methods used at a slightly lower count, they would try to keep on using the lower staple by altering their production methods somewhat.

It follows that diagrams of the extra cost imposed on ring spinning because of the need for a longer staple would in fact differ somewhat from Figures 3 and 4.[1]

Nevertheless, some conclusions of relevance to this paper can be drawn from the information available. First, the differential in cotton costs between rings and mules for warp yarn

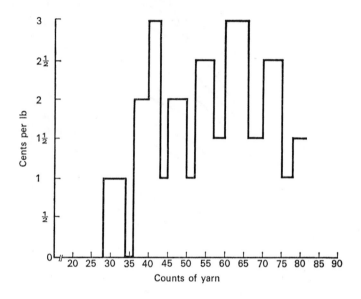

Figure 3

EXTRA COST OF THE LONGER STAPLE
NEEDED FOR THE RING SPINNING OF WARP (TWIST)

probably starts to appear at a count around 28 and then increases to a peak somewhat below three cents per pound, probably in the vicinity of 45 or 50. In all probability, the difference reaches two cents in the low 40's. Secondly, the cost differential does not drop significantly below two cents again, at least not in the range shown in Figure 3. As for weft, the cost differential starts around a count of 35 and rises to more than two cents in the 50's. It probably reaches two cents in the upper

[1] A more detailed discussion of this question can be found in the author's longer paper which is available on request.

40's. The differential then stays at least as high as one and one-half cents for higher counts.

The discussion so far has only dealt with counts below 100. This is the area relevant for Great Britain. In the United States, however, the only count range where mules appear to have been installed was above 100. The question thus arises as to whether cotton price differentials can explain at least the very partial return to mules at very high counts.

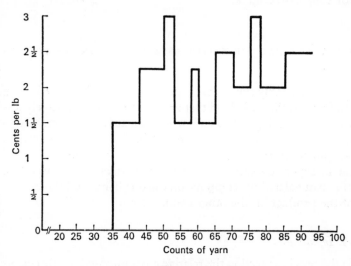

Figure 4

EXTRA COST OF THE LONGER STAPLE
NEEDED FOR THE RING SPINNING OF WEFT

On the whole, it can be expected that price differentials of two or even three cents continued out well beyond a count of 90. This continued gap resulted from the need to resort to Egyptian and 'regular' Sea Island cotton for ring twist in the 80 to 100 range.[1] Above that range, however, it eventually became necessary to use 'Best Sea Island' cotton. In view of the very large difference in the count that could be spun by the two methods at these high counts,[2] the differential must at some point have been between using the Best Sea Island on rings or

[1] Todd, op. cit., p. 17.
[2] Ibid., and Winterbottom, op. cit., p. 52.

distinctly inferior types of cotton on mules. The cost differential can be estimated to be five or six cents per pound and sometimes even more. Once the count is high enough, however, even the mule would require Best Sea Island. Once this happened, the cost differential would be very much reduced, at least if the ring was physically capable of spinning such extremely fine yarn. This evidence appears to be consistent with rational manufacturers installing both mules and rings for very high counts.

RELATIVE FACTOR COSTS AS A FUNCTION OF COUNTS SPUN

I have repeatedly stated that expert opinion in the period being studied unanimously held that ring spinning was relatively less well suited to high than to low counts. It is now necessary to look at this proposition in more detail. Changes in the relative advantage of ring and mule spinning as the count increased can be expected to result from changes in the relative cost of the raw material and other inputs, on the one hand, and the quality of the product on the other hand.

A. Input Costs

In the previous section the relative costs imposed by the cotton needed for ring and mule spinning was studied as a function of the count of the yarn spun. It appears that this cost difference did increase with the count, at least in moving from low to medium and high counts.

The other important inputs to be examined are labour and capital. It is quite clear from all contemporary evidence that labour input per pound of yarn increased faster on rings than on mules as the count increased. Even with a constant capital-labour ratio, this would also imply that the capital cost of ring spinning increased faster than that of mule spinning. In fact, however, the capital–labour ratio increased faster in ring than in mule spinning.[1] In ring, unlike mule, spinning, the number of spindles per operative increased as the count increased. The

---

[1] Jewkes and Gray, op. cit., p. 121.

difference in capital costs thus increased even faster than implied by the changing ratio of spinning labour input.

Before jumping to the conclusion that this evidence proves that the cost of ring spinning increased faster than the cost of mule spinning, it must be remembered that mule spinners were more expensive than ring spinners. Thus, an equal percentage increase in the number of spinners' hours per pound of yarn would increase the saving per pound of yarn to be derived from using rings rather than mules.

While a good deal of vague information on the relationship between the count of yarn spun and labour input is available for the pre-First World War period, careful studies of this relationship were carried out only in a later period. Two such studies, one based on interwar conditions and one based on 1949 conditions, are available.[1] Happily, the two studies present almost identical results with regard to the relationship of output per man hour in spinning and the count of yarn spun. I therefore used the results of these studies to calculate cost differentials.[2]

In addition to an estimate of output per man hour, I also needed an estimate of changes in the spindles per man ratio in ring spinning. I obtained an estimate of the latter from the structure of piece rates in the British Universal Ring Spinning List 1912.[3] This structure was specifically designed to reflect the fact that spinners working on higher counts were able to tend more spindles. On the basis of this information, I calculated the saving per pound of yarn in spinning labour and capital charges that resulted at various counts from using rings instead of mules. Assuming a waste rate of 5 to 10 per cent, the

---

[1] See British Ministry of Production, *Report of the Cotton Textile Mission to the United States of America* (London: His Majesty's Stationery Office, 1944), and Productivity Team, *Cotton Spinning* (London: Anglo-American Council on Productivity, 1950).

[2] It is, of course, unfortunate that these studies do not refer to the exact period under study. Encouragement, however, can be taken from the fact that no noticeable change occurred between the 1930s and 1949. More important, virtually all the mules studied, and the great majority of the rings were in fact installed before the First World War. If there is any bias in using these past period studies, it is probably in underestimating the labour required on high count rings. Such a bias might be expected because technical change on the ring generally is credited with making it effective at higher and higher counts.

[3] Jewkes and Gray, op. cit., p. 121.

saving per pound of cotton used would be 5 to 10 per cent less than the results shown in Table 1.[1]

TABLE 1 *Cost Differentials*

Difference in Labour and Capital Costs in United States Cents							
Count	40	50	60	70	80	90	100
Great Britain	1·6	1·7	1·8	1·8	1·8	1·6	1·5
United States	2·4	2·6	2·9	2·9	3·0	3·0	2·9
Count			110	120			
Great Britain			1·2	0·8			
United States			2·7	2·3			

B. *Quality Differentials*

There now remains the question of the quality of ring versus mule yarn. There is a great deal of talk, especially in British writings, concerning the superior quality of mule yarn. Clearly, however, British comments on this subject are of questionable value. After all, they can be counted on to claim superiority for the product on which they concentrated. I have also been unable to find any data that show a price differential between yarns of the same count, made of cotton of the same quality (here defined to exclude staple length), differing only in the method of production.

Nevertheless, there do seem to have been some differences between the two types of yarn. Thus, Copeland, an American, remarks that: 'Mule yarn, however, is superior . . .', and 'the harder ring-spun yarn is better adapted for warp than for weft.'[2]

---

[1] This calculation is based on the assumption that the capital costs of mules and rings were the same at a count of 40. Account has been taken of the somewhat greater speed of rings in the United States as compared with Great Britain. The generally accepted figure of 10 per cent for loss, depreciation, and upkeep of machinery (see Winterbottom, op. cit., p. 271) is used. In addition, the interest cost of the money invested is set at 10 per cent. While the cost differential at a count of 40 is independent of these percentages, the cost differentials at higher counts would be higher if lower interest rates were used and lower if higher interest rates were used.

There is some reason to believe that the rate of growth of the capital differential is underestimated. This bias results because high-count mules have shorter 'draws' than do low-count mules and, therefore, occupy less space than do low-count mules. This is not the case with rings.

[2] Copeland, op. cit., p. 68.

My general conclusion in this matter is that mule yarn probably did have some superior qualities. It is not clear whether this advantage became greater in a technical sense as the count increased. What is clear, however, is that the *importance* of this difference increased with the count. For the low-quality cloth usually made with low-count yarns, this minor difference in the yarn probably did not matter much. As the quality of the cloth increased with the count, however, differences in the yarn undoubtedly took on added importance. Quality differences thus probably did make the ring somewhat less well adapted to high than low counts as well as more suited to warp than weft. This difference in suitability for warp and weft probably played a role in what appears to have been the greater staying power of old American mules in weft as opposed to warp spinning.

COSTS AND BENEFITS

Having estimated the various costs and benefits involved in choosing between mules and rings, it is now time to evaluate the results.[1]

Taking account of all these different costs (including labour, fuel transportation, and capital) with the single exception of cotton, it seems that in Great Britain the saving per pound of cotton spun on rings rather than mules was about 1·5 cents for warp and 1·2 cents for weft at a count of 40. This saving rose to about 1·7 and 1·4 cents, respectively, at a count around 70 and then declined slightly. On the other hand, the increase in cost due to the longer staples required by ring as opposed to mule spinning for twist (i.e. warp) rose from zero at counts below 28 to about 2 cents around 40. It then remained at that level. For weft, the differential probably reached 2 cents close to 50. It thus appears that in Great Britain rings were preferable for warp production up to a count perhaps a little below 40, while for weft they were probably to be preferred even for counts in the low 40's. In cases where the spinner contemplated

[1] The longer version of this paper also contains prima facie evidence that French and German manufacturers were rational in choosing between rings and mules and that British and American manufacturers were rational in their replacement policies.

using low-quality cotton, rings may have been better even at slightly higher counts. It does not appear that rings ever became profitable again at higher counts. This conclusion is reinforced by the fact that a growing effective quality differential probably worked against ring yarn at higher counts.

When these results are compared with the actual behaviour of British manufacturers, they appear to have behaved in a rational manner. At the very least, these results should throw the burden of proof onto those who maintain that the British were irrational in their choice between rings and mules.

In the case of the United States, the cost advantage per pound of yarn was over 2 cents at a count of 40 and then rose to around 3 cents. In view of these results, it is understandable that ring spinning was used for much higher counts in the United States than in Great Britain. At the same time, the price differential is down to 2·3 cents per pound at a count of 120 and heading lower. In view of this fact and the high price differentials encountered for very long staple cotton, it is also not surprising that some mules were being installed to spin very fine yarn. Indeed, my reaction is that surprisingly few new mules were installed in the United States. Although the quantitative data are not strong enough to prove the point, I suspect there may have been some substance to the many comments by contemporary observers that employer dislike of unions caused them to avoid mule spinning.

5 The Market and the Development of the Mechanical Engineering Industries in Britain, 1860–1914

S. B. SAUL

[This article was first published in the *Economic History Review*, 2nd series, Vol. XX (1967).]

I

This article has been written in a mood of dissatisfaction with recent general interpretations of the performance of British industry in the half-century prior to the First World War, and in the belief that the time has come to turn away from generalizations based on a limited range of secondary evidence and to find out more precisely just what was happening.[1] We still lack detailed studies of most industries, and this is particularly true of that complex and heterogeneous sector, mechanical engineering. There has been virtually no analysis at all of engineering technology in Britain. The development of the techniques of standardization and interchangeable manufacture, for example, has been studied almost entirely from the American point of view. It is tacitly assumed that in this area of engineering technology British industry lagged particularly badly, a view which, to my mind, is seriously misleading. Much has been written too in a very general way about the quality of entrepreneurship and its role in bringing about a declining rate of growth in Britain. Again the evidence so far adduced is remarkably thin. Take the third generation argument, for example, derived apparently from the steel industry, Marshalls of Leeds, and little else. In engineering it would not be difficult to point to opposite examples – to the self-made, semi-literate genius giving his son the conventional but thorough

---

[1] This article is a revised version of a paper presented to the Economic History Conference at Manchester in April 1966.

practical engineering education of the time, and the third generation being sent to a university engineering school.[1] But in any case, few firms were born before 1850 and the third generation was not reached until after 1914. In any case, too, as I shall show later in the course of this paper, entry was relatively easy to many engineering trades, and they were subject to constant rejuvenation in that way. One answer to the question of the role of the individual entrepreneur is more study of the origins and training of businessmen. Another, which is attempted here, is to try to determine the degree to which explanations in terms of entrepreneurial weakness are required, by studying the environment in which the entrepreneur operated. This article lays particular stress on one determinant of entrepreneurial behaviour – the nature of the market. Professor Habakkuk has already discussed this largely from the point of view of a general deficiency of demand, whereas here a more sectoral and institutional approach is adopted.[2] The intention is to show that the market is important in explaining some of the shortcomings of British engineering before the war, and that, to some degree at least, it was beyond the power of the individual firm to alter these objective environmental conditions.

Be that as it may, it is also important to remember that many engineering trades showed remarkable resilience and vigour in this period. Table 1 indicates the relative importance of these trades in 1907.

The neglect of textile engineering is puzzling: it hardly gets a mention anywhere, not even by Clapham. Yet in 1907 it was the largest single branch of engineering and an overwhelmingly dominant force in world trade. In 1913 about 40,000 men were employed, over three-quarters of them by six very large manufacturers of cotton machinery in Lancashire – the details are given in Table 2.[3]

It was, too, an industry apparently kept well on its toes by the infusion of new blood – Howard and Bullough and Brooks and

---

[1] See, for example, Rachel E. Waterhouse, *A Hundred Years of Engineering Craftsmanship* (Tangyes Ltd, 1957), and Eric N. Simons, *Jenkins of Rotherham* (Robert Jenkins Ltd, 1956).

[2] H. J. Habakkuk, *American and British Technology in the Nineteenth Century* (Cambridge, 1962), pp. 185–8.

[3] *The Times Engineering Supplement*, 9 July 1913, p. 9.

Doxey, founded only in the 1850s, were in many ways the technological and commercial leaders after 1870. In 1894, for example, Howard and Bullough set up an American subsidiary,

TABLE 1 *Gross Output of Selected Engineering Industries in Britain in 1907 (£m.)*

Textile Machinery		13
Railway Locomotives:		
Private builders	4·5	
Railway Company (construction and repair)	7·9	
	——	12·4
Steam-engines (excluding locomotives and agricultural steam-engines)		6·9
Boilers		4·1
Machine Tools		2·9
Agricultural Machinery	1·1	
Agricultural Steam-engines	1·3	
	——	2·4
Cycles, Motor Cycles, and Parts		5·6
Motor Vehicles and Parts		5·2

Source: First Census of Production, 1907.

and their reputation and their ability to offer a full line of equipment for a spinning mill made them fearsome competitors in that market.[1] The large home market was a clear advantage,

TABLE 2 *Principal Makers of Textile Machinery to 1914*

Platt Brothers, Oldham	Founded 1821	Employment 1913: 12,000 (incl. about 1,000 in three collieries)
Howard and Bullough, Accrington	Founded 1853	Employment 1914: 6,000
Dobson and Barlow, Bolton	Founded 1790	Employment 1908: 4,000
Brooks and Doxey, Manchester	Founded 1859	Employment 1913: 4,000
John Hetherington, Manchester	Founded 1837	Employment 1913: 4,000
Asa Lees, Oldham	Founded 1790s	Employment 1913: 3,000

Sources: D. A. Farnie, *The English Cotton Industry, 1850–1896* (Manchester M.A. Thesis, 1953), chapter 4; *The Times Engineering Supplement*, 9 July 1913, p. 9; and information from the companies.

but by 1907, 45 per cent of the industry's output was exported, and only a quarter of that went to the British Empire. After India, the main markets were in Europe – Germany, Russia,

[1] J. R. Navin, *The Whitin Machine Works since 1831* (Cambridge, Mass., 1950), p. 241.

France, and Belgium. By 1914 the U.S.A. was the only area in the world not dependent on Britain for a major part of its textile machinery. Even there Keighley firms monopolized the market for worsted machinery.[1] *The Times* suggested in 1913 that, except for those in the United States, Austria, and Switzerland, the world's cotton spindles had been mostly bought in Britain.[2] It is easy to lose perspective because the American industry is well analysed in two excellent histories of the Lowell and Whitin works, but Platts of Oldham, of which there is no history at all, had an output in 1914 equal to that of the whole American industry put together. Each week they turned out 20 mules, over 100 ring frames, 320 looms, 80 carding engines, as well as many other ancillary machines.[3] In some parts of the world Platts's machinery was standard for all cotton mills. Unlike the American industry, which had grown so customer-conscious in its fight for the southern mill orders that it was prepared to grant all kinds of design concessions to individual millmen, the British were in a position to insist on standardized products. Chinese spinners, for example, stated that they preferred British to American machinery as the latter changed so often that there was great difficulty in securing new parts for replacement.[4]

Gibb describes American development to 1900 as the combined result of original thinking and copying from the British, while Navin writes of the traditional American way of catching up in this technology – purchasing British machines through a third party, copying them minutely and advertising them as such. This was how Lowells attempted to break the Keighley monopoly of worsted drawing and spinning machinery in 1898, for example, though with little success.[5] On the other hand, two of the main technological advances of the later nineteenth century were pioneered in the United States – ring spinning and the Northrop loom. As for ring frames, the great productive

[1] In the main cotton textile centre of China, even in 1918, 94 per cent of the spindles had been supplied by Britain. In 1913, 80 per cent of total Chinese imports of textile machinery came from Britain. See H. D. Fong, *Cotton Industry and Trade in China* (Tientsin, 1932) Vol. I, 79–80.

[2] *The Times*, loc. cit.

[3] J. R. Navin, op. cit., p. 324, and information from Platts Ltd.

[4] H. D. Fong, op. cit., p. 79.

[5] G. S. Gibb, *The Saco-Lowell Shops* (Cambridge, Mass., 1950), p. 260, and J. R. Navin, op. cit., pp. 375 and 391.

power of the British industry enabled it to improve on ring spinning techniques and sell in the American market itself despite the freight and duty.[1] American-made ring frames were not exported at all before 1914. The Rabbeth spindle, which made ring spinning a commercial success, was introduced into England by Howard and Bullough in 1878. In seven years they sold 1¾ million spindles – less than half of them at home – and another 2 million in the next four years. Employment boomed from 350 in 1870 to 2,000 in 1892. Platts reached a similar level of output in the early 1890s. Brooks and Doxey matured under the same influence: by 1892 they were employing 2,000 people and making 2,000 ring spindles a week as well as preparatory machinery and some machine tools. Tweedale and Smalley of Rochdale, only founded in 1892, were already employing 1,700 workers, largely on ring frames, in 1900.[2] All this indicates an industry which had fully retained its virility and competitiveness. Nor should it be forgotten that extensive improvements were being made for most of this period to the mule. Indeed, one technical expert has suggested that the 1870s and 1880s were outstandingly productive in this respect.[3]

Turning now to the more specific question of standardization and interchangeability, apart from the Portsmouth block-making machinery of the beginning of the nineteenth century, almost all the credit for developing these techniques has gone to American engineers. Undoubtedly their contribution was enormous through the creation of new types of machinery to machine small parts accurately in mass – turret lathes, milling and grinding machines – but what of the British role? In so far as Britain's pre-eminence lay in heavy capital goods, the opportunities for mass production were limited, but once specialist manufacture became the rule, and where the market existed, or could be created, to warrant investment in using these techniques, there is little evidence of technological shortcomings. The major firms in the textile machinery trade were specialists and in no sense general engineering firms,

[1] D. A. Farnie, op. cit., p. 64.
[2] Ibid., pp. 66–8; *The Century's Progress. Lancashire* (1892), p. 80, and information from the companies.
[3] *Textile Manufacturer*, December 1925, p. 49.

though all the big companies produced a wide range of textile machines. Spindles were certainly made interchangeable from at least the middle of the nineteenth century, and by the end of the century were being turned out by individual firms in very large quantities – in batches of up to 100,000. It appears, however, that right up to 1914 they were not produced on modern machinery but were made by extremely skilled and experienced workmen using a grindstone and achieving an accuracy of up to 0·005 in in places, on what were small diameters, with the use of emery cloth. They were, of course, using the 'go, no go' gauges required for interchangeability rather than seeking dead accuracy.[1] It seems primitive, but it is by no means clear that it was economically unsound, especially when one reads of the difficulties experienced by American firms in matching with machinery 'the superb work of the English fitters'.[2] As for the rest, textile machinery parts were not altogether truly interchangeable without a moderate amount of fitting. By and large, the technology was one of skilled repetitive work using conventional machine tools combined with considerable numbers of special-purpose tools designed and made by the textile machinery firms themselves. However distinguished their role in world textile engineering generally, and despite the size of their output, their machining techniques were a mixture of the conventional – not to say old-fashioned – and the highly specialized. The industry did little after 1850 to advance engineering technology in general, nor was it an important centre for general engineering training.

More interesting, perhaps, is the production of locomotives, the second largest of the engineering trades and one where Britain's performance was not quite as satisfactory. Standardization and interchangeability were certainly not confined to, or originated by, American builders. Sharp Roberts built interchangeably to templates and gauges at a very early date, and theirs was one of the first works in the world to use the system.[3] To ensure interchangeability for his first series of standard locomotives for the Great Western in the early 1840s,

---

[1] For information on manufacturing techniques I am most grateful to J. S. Taylor of the Department of Textile Technology, Manchester College of Science and Technology, and to H. Catling of the Shirley Institute.

[2] G. S. Gibb, op. cit., p. 345.

[3] J. Roe, *English and American Tool Builders* (New Haven, 1916), p. 62.

Daniel Gooch sent out full specifications and templates to the builders employed, this being the first attempt at standardization on an extensive scale for an individual railway.

> These drawings I took great pains with, giving every detail much thought and consideration ... When I had completed the drawings I had them lithographed and specifications printed, and thin iron templates for those parts it was essential should be interchangeable, and these were supplied to the various engine builders with whom contracts were made. One hundred and forty-two engines were let, and all the makers did their work well.[1]

How accurate all this was, and how much hand fitting was required in assembly is not known, but the principle was there. It reached its peak when John Ramsbottom was chief engineer at Crewe. Having been trained at Sharp Roberts, he had absorbed their technological ideas. Between 1858 and 1872, 857 goods engines were constructed at Crewe for the London and North Western Railway, and another 86 between 1871 and 1874 for the Lancashire and Yorkshire Railway – 943 locomotives identical in almost all respects, a record which was not subsequently surpassed.[2]

But if all this was so, why was it that an American firm such as Baldwins was able to take the technique to its logical conclusion, building standard engines for stock on a very large scale, employing 19,000 men in 1907, whereas the largest company in Britain employed less than half that number and the largest single private establishment 3,400 (see, generally, Table 3)?[3]

The answer lies largely, I think, in a study of the market. Unlike most other countries – Germany and the U.S.A. in particular – in Britain the major railway companies made and repaired their own engines in what were some of the largest engineering

---

[1] *Diaries of Sir Daniel Gooch, Bart.* (1892), pp. 40–1.
[2] E. L. Ahrons, *The British Steam Locomotive, 1825–1925* (London, 1927), p. 123.
[3] *History of the Baldwin Locomotive Works, 1831–1907* (Philadelphia, 1907), p. 104. German locomotive works were comparable with the British in size. Borsig in 1907 produced 300 locomotives; the North British 485 locomotives. See U.S. Bureau of Manufactures, *The Machine Tool Trade in Germany, France, Switzerland, and the U.K.* (Washington, 1909), p. 34. North British list supplied by the Stephenson Locomotive Society.

establishments in the country. Apart from the eight major works listed in Table 3, there were at least as many again building new engines in 1914. These company workshops were private empires largely isolated from the market. The strong individuality of the chief engineers meant that there was little

TABLE   3 *Employment in the Main Locomotive-Manufacturing Works in Britain*

Railway workshops. c. 1914			
Swindon	14,000	St Rollox	3,000 (1900)
Crewe	7,000	Cowlairs	2,000 (1895)
Stratford	7,000		
Derby	5,000		
Doncaster	4,000		
Horwich	4,000		
Private constructors: 1907			
North British	8,000	(in 1902, prior to amalgamation, employment in the three firms was:	
		Nielsen	3,400
		Dubs	2,400
		Sharp Stewart	1,700)
Beyer Peacock	2,700		
Kitsons	2,000		
Vulcan Foundry	1,700		
R. Stephenson	1,200		
Nasmyth Wilson	500		

Sources: Private constructors: *Glasgow Herald Annual Trade Review*. Railway workshops: from technical journals and the companies. Details available from the author.

interchange of information or uniformity of practice. Some were brilliant designers, some definitely were not. F. W. Webb, for example, chief engineer at Crewe from 1871 to 1903, was a man of the most extreme dogmatism and prejudice, and suffered from increasingly serious bouts of insanity during the last years of his career.[1] The best works were very advanced in equipment and practice. Those at Horwich were described in 1891 soon after their opening as possessing 'some of the finest milling plant we have yet seen'. Great strides were made there and elsewhere in standardizing rolling stock and ensuring interchangeability of cylinders, valves, axle-boxes, springs,

[1] Unpublished article by Dr W. H. Chaloner.

wheels, and so on.[1] Milling was introduced into the Stratford works of the Great Eastern Railway in 1878.[2] The Gateshead works were reorganized in 1883 and 1884, and with the arrival of T. W. Worsdell as chief engineer the following year, there was a marked increase in the use of milling machines for all kinds of heavy work and for finishing processes.[3] G. J. Churchward took charge of Swindon in 1902. With a strong functional outlook and little interest in aesthetics, he launched a major programme of re-equipment through standard types of engine with many parts common to all – piston rods, crossheads, valves, axles, springs, and a standard bogie too.[4]

However, the railway companies were unable to capitalize on all this and manufacture locomotives for others – the sale of engines made at Crewe to the Lancashire and Yorkshire Railway resulted in the private builders successfully applying for an injunction in 1876.[5] The private builders were relatively few in number in 1914 – all the significant makers of large locomotives are shown in Table 3. The best private works were as well equipped and organized as those of the railway companies. Dubs's machine shop was described in 1887 as one of the finest in Britain, using milling techniques extensively and 'the result of this ensures such absolute uniformity of dimensions that the firm can at once duplicate any part of any engine ever made by them', and the organization such that 'no man ever leaves his machine to grind his cutting tools, there being a squad of men who do nothing else all the year round'.[6] A visitor to the Beyer Peacock locomotive works in 1886 reported that 'all holes are drilled through templates: in fact everything is machined to templates so that corresponding parts are interchangeable'.[7] The amalgamation of three companies in the Springburn district of Glasgow to form the North British Company in 1902 was certainly in part a response to overseas competition, but the prime difficulty remained that the

---

[1] *The Engineer*, 7 August 1891, and *Transactions of the Institution of Mechanical Engineers*, 1909, p. 561. (Subsequently referred to as *Mechanical Engineers*.)

[2] *Mechanical Engineers*, 1896, p. 545.

[3] *Round the Works of our Great Railways* (1893), p. 99.

[4] H. Holcroft, *Outline of Great Western Railway Practice* (1957), pp. 81–7.

[5] O. S. Nock, *The Premier Line* (1952), p. 53.

[6] *Album of Arts and Industries of Great Britain* (1887), p. 321.

[7] *The Engineer*, 29 October 1886, p. 341.

market structure gave the private builders little opportunity of building up quantity production of relatively cheap general-purpose engines. The home market was limited for the reasons I have mentioned. Of 937 engines sold from 1872 to 1914, Nasmyth Wilson placed only 7 per cent on the home market. The Vulcan Foundry sent 70 per cent overseas from 1894 to 1905, and from 1906 to 1914 sold at home only 17 of 948 engines produced. Compare this with the experience of Baldwins who in 1906 sold 292 engines overseas and 2,374 at home, or the Hanover Locomotive Works which from 1846 to 1907 sold only one-fifth of the 5,000 engines turned out overseas.[1] When they did give orders, the home railways often refused to take stock patterns, but insisted on what suited them and fitted best into their own repair and maintenance patterns. The private builders got their bread and butter orders overseas, and these came from railways in the Empire and South America run by British railway engineers. These men, too, rarely ordered from stock, but designed in detail precisely what engines they wanted. The chairman of the North British Company told the Tariff Commission in 1905 'it is practically impossible to make parts to stock: we never know what the next orders will be'. And he added ruefully, 'in our work as contractors' – note the term – 'there is not much scope for men of an inventive turn of mind'.[2] The builders complained bitterly too about the army of inspectors who invaded their shops whenever orders involving the Crown Agents were being fulfilled. They also argued that their costs were raised by the consulting engineers' practice of forcing them to obtain bought-out parts from specified firms who took advantage of their position to inflate their prices. This institutional framework therefore resulted not only in the market being created in an entirely different manner in Britain and America, but in the building of entirely different technical products. The British engine was a high-quality product with a long life, relatively expensive and difficult to service, closely tailored to the needs of the line for which it had been specifically designed. It was not necessarily very suitable for other purposes, and was particularly unhappy running on poorly

[1] Lists of the Stephenson Locomotive Society; *History of Baldwin Locomotive Works*, p. 100; U.S. Bureau of Manufactures, op. cit., p. 80.

[2] *Tariff Commission*, Vol. IV (1909), 579.

constructed tracks.[1] For their type of engine the British makers were unsurpassed. Their works were well equipped: standardization within the runs they were given was a commonplace, but by the last decade of the nineteenth century not only had they no sound market basis for producing standard engines cheaply for stock, but the engineering traditions which the market had built into them would have made it nigh on impossible for them to adopt the practice anyway.

Knowledge of production techniques in the manufacture of steam-engines – stationary and marine – is scarce, but when the market was right, interchangeability was aimed for and achieved from an early date. An outstanding example of interchangeable marine-engine construction arose out of a rush Admiralty order for shallow-draught gunboats during the Crimean War. One hundred and fifty sets of engines were successfully built by Penns and Maudslay to that principle – the first example of mass production in marine engineering.[2] There were, of course, innumerable makers of small stationary engines, and in general at the international exhibitions of the 1860s and 1870s their products came under severe criticism for their poor fuel economy, something which might not matter too much at home but made them unsaleable overseas. However, already in those decades one or two outstanding makers were striking out into new directions. Hick Hargreaves of Bolton sponsored the American Corliss engine here, and in 1867 began making it simply and to a standardized design.[3] In the late 1860s James Tangye had a huge success when he brought out a simple engine manufactured interchangeably and built for stock in large numbers.[4] He created a separate machine-tool section to build the tools for the purpose, and then went on to sell such machines to industry generally. Even more significant was the work of Peter Willans who introduced the principle of standardization between engines of different sizes. The high-

---

[1] See F. Jukes, 'Passenger Locomotives, British *v.* American', *Railway and Locomotive Historical Society*, X (1959), 56; *The Engineer*, April 1899, p. 341; *The Times*, 1 June 1900, p. 13.

[2] G. A. Osborn, 'The Crimean Gunboats', *The Mariner's Mirror*, LI (1965), 106–8, and information from J. Foster Petree.

[3] *A History of Technology*, ed. C. Singer et al. (Oxford, 1958), Vol. V, 132.

[4] Rachel Waterhouse, op. cit., p. 31, and *Mechanical Engineers*, 1913, p. 640 and 1921, p. 548.

F

pressure cylinder of one size was the same as the low-pressure cylinder of the next smaller size but one, and so forth: interchangeable cylinders, pistons, rings, valves, were made in quantity for stock.[1] An American engineer commented in 1897: 'I do not know of an American shop building engines which carries the principle so far.'[2] James C. Peache went to Willans's works at Thames Ditton to take charge of the system of interchangeable manufacture in 1885. In 1891 he invented his own high-speed engine to the same pattern, and it was later extensively manufactured by Davey Paxman. Savorys of Birmingham made small very high-speed marine engines to the same principle too.[3] These were pioneers, but their output became large, their fame great, their works became models and many came to see and learn there. So far had things gone that the Admiralty had six cruisers built with interchangeable main and auxiliary engines in 1903 – the first navy to do this – using as chief contractor Hawthorn Leslie who had fifteen years' experience of the work with smaller naval craft.[4]

A related sector of the engineering industry where the market complicated the position was the manufacture of agricultural machinery. Several large firms in the eastern counties achieved striking success both at home and overseas in those lines most in keeping with British engineering traditions – portable steam-engines, threshers, ploughs, and, later, oil-engines for farm use. Their business expanded rapidly after mid-century. Marshalls of Gainsborough, the most important, founded in 1856, employed 550 by 1870, 2,000 in 1892, and 5,000 in 1913. Clayton and Shuttleworth established a general iron foundry in Lincoln in 1842, employed 1,400 in 1885, and 2,300 in 1907. Ransomes of Ipswich, a much older firm dating back to the eighteenth century, employed 1,500 in the mid-1880s and grew fast largely on the basis of exports during the 1890s to reach 2,500 in 1911. Rustons of Lincoln, founded in 1857, employed 5,200 in 1911. Less than half of their work-force was engaged on agricultural machinery, though perhaps a

---

[1] *History of Technology*, Vol. V, 136, and K. W. Willans, 'Peter Paul Willans', *Transactions of the Newcomen Society*, XXVIII (1951–3).

[2] *American Machinist*, 10 June 1897, p. 425.

[3] *Mechanical Engineers*, 1931, p. 753, and information from R. H. Clark.

[4] Fred T. Jane, *Fighting Ships, 1906–07* (1907), pp. 444–7, and *Engineering*, 4 May 1906.

third of their oil-engines were for farm use.[1] Howards, plough makers of Bedford, and Fowlers with steam-plough equipment, were also major exporters. These firms dominated the international exhibitions of the late nineteenth century in their own specialities just as much as the Americans did in theirs. Clayton and Shuttleworth were described at the Vienna exhibition as having machines all over the Austro-Hungarian Empire, and in fact had a factory in Vienna, set up in 1857 and soon employing 700 men. They had another at Pesth where Robey of Lincoln also had a works.[2] It was fortunate that they did too, for tariffs and government subsidies to the State manufactory of portable engines had largely eliminated imports by 1900. Up to 1914 Russia was Ransomes' big market and Marshalls' and Garretts' too, and all lost heavily on account of money lodged with Russian banks when the revolution broke out.[3] The selling mechanism was particularly highly developed. The organization included a staff of travellers, mechanics, spare-part depots, and repair shops, and in most cases a complementary organization which bought grain from the farmer, thereby allowing a purchase of machinery and sale of produce to be carried out conveniently in one transaction.[4] So much, in this field at least, for the poor British reputation for salesmanship. The home trade was negligible: output of steam threshing sets in 1913 was above 100 a week of which, on average, only two were sold at home.[5] By weight, fully half the exports in 1913 went to Europe, an unusually high proportion. Between two-thirds and three-quarters of the total output of agricultural machinery was exported – a high proportion, especially when one recalls that very few binders or reapers went overseas. The works were apparently modern and well equipped, though considering the range of equipment all produced and the large size of much of it, they were in no

[1] *Mechanical Engineers*, 1890, pp. 441, 450, 554; *Ransomes' 'Royal' Records, 1789–1939* (Ransomes, Sims, and Jefferies Ltd, 1939); B. Newman, *One Hundred Years of Good Company* (Ruston and Hornsby Ltd, 1957), pp. 21–35, 68, 80; and information from the companies.

[2] *The Engineer*, 18 October 1867.

[3] Ipswich Engineering Society, *History of Engineering in Ipswich* (Ipswich, 1950), p. 61; R. A. Whitehead, *Garretts of Leiston* (1964), p. 193, and information from the companies.

[4] Committee on Industry and Trade, *Survey of Industries* (1926), IV, 164.

[5] *Report on the Engineering Trades after the War*, P.P., 1918, Vol. XIII, para. 64.

sense engaged in mass production – at their peak Clayton and Shuttleworth produced 25 threshing sets a week.[1] Marshalls, the largest, as early as 1885 were extensive users of milling machines, twist drills, and the like; Clayton and Shuttleworth employed as works manager to 1914 H. F. L. Orcutt, who had helped plan Loewe's machine-tool works in Berlin and was a constant advocate of American production methods.[2] Ransomes' factory in 1905 had a great deal of American machinery, and J. E. Ransome himself commented that 'firms with antiquated tools generally go down. It is a very bad sign when a works is full of old tools; it does not pay to keep them.' When questioned on interchangeability he was somewhat contemptuous. The firm had been using standard interchangeable spare parts since before he was born: they could not service machinery supplied to farmers in remote parts of the world in any other way.[3]

But of course British agricultural machine production failed in the one area where the American mass interchangeable techniques were so relevant – in reaping and binding machinery. In 1900 the entire British output of harvesters was not a tenth of that of McCormick.[4] It is true that during the 1870s, when the Americans seized the whole of world trade, British makers failed to appreciate what was happening until too late. They were too confident that American machines were too light and applicable only to the United States home market, whereas the machinery was in fact quickly and intelligently adapted to local requirements. But the market was again a difficult one. The new machinery was economic only on large, level fields with no boggy patches or land-fast stones. The lower labour costs, the mixed pattern of farming, greatly intensified after 1870 with the drastic fall of the arable acreage, all made it extremely difficult to establish a strong home market and to cut down the American lead which was itself in part a reflection of her favourable home market conditions. To all this had to be added the conservatism of the British farmer himself. Discussing this point, an official report noted the suspicion

[1] *Engineering*, 21 June 1907.
[2] Ibid., 7 August 1885 and 21 June 1907.
[3] *Tariff Commission*, Vol. V, 668 and 677.
[4] *The Times*, 7 June 1900.

and often undisguised hostility of farmers to all innovations and commented 'manufacturers have had to contend with much inertia and prejudice in bringing their appliances to the notice of the farming public'.[1]

In their own specialities the British firms carried out with marked success what was largely an export trade, producing their equipment by up-to-date methods and selling it vigorously. The market, however, played a significant part in limiting the extension of Britain's output of agricultural machinery. This is not to argue that it was the only factor: a report on the Paris Exhibition of 1900 stated 'It is useless to deny that amongst the jurors there was a feeling that amongst British agricultural implement makers there was a certain want of progress and they were too much inclined to rely on their undoubted triumphs of the past.'[2] This comment, however, almost certainly applied to the industry's inability to move into the harvester field, rather than to any weaknesses in the traditional fields. It may have been this sense of despair which caused an observer to comment on the number of American machines shown at the Maidstone Agricultural Exhibition of 1899 that 'some of our leading firms were becoming implement agents rather than implement sellers'.[3] We shall see later that the work of importing agents had a more stimulating effect on the machine-tool industry in the more favourable market conditions obtaining for that sector after 1890.

II

So far there has emerged a pattern of the classic engineering industries continuing to enjoy marked commercial and technological progress, moving into new fields such as the manufacture of ring frames, adopting new techniques of standardization, but with their success being limited here and there by

[1] *Report on Agricultural Implements and Machinery under the Profiteering Acts 1919 and 1920*, P.P., 1921, XVI, 4.
[2] *Report of Her Majesty's Commissioners for the Paris Universal Exhibition 1900*, P.P., 1901, XXXI, 157. Britain lost most of the world's plough trade to German, Canadian, and American firms largely on account of her unwillingness to produce anything approaching a standardized article. Fowlers, for example, turned out ploughs with 200 different mould boards and 58 plough breasts. See A. Pepper, *Retrospect of over 50 years with John Fowler and Co. (Leeds) Ltd* (1946), p. 8.
[3] *Journal of the Royal Agricultural Society of England*, 3rd ser. X (1899), 552.

peculiar market problems. But now to go a step farther. The
most influential branch of all engineering production is the
machine-tool industry, for it is there that new skills and techniques are acquired and diffused. Several writers have attri-

TABLE 4 *Principal Specialist Machine-Tool Makers in Britain, 1840–1914*

Date*	Name	Location	Employment c. 1913 except where shown
1842	Buckton	Leeds	400
1842	Muir	Manchester	400 in 1894
1852	Hulse	Manchester	300 in 1888
1853	Craven	Manchester	1,000
1856	Greenwood and Batley	Leeds	2,000 in 1903
1857	Kendall and Gent	Manchester	400
1859	Smith and Coventry	Manchester	500 in 1894
1864	Cunliffe and Croom	Manchester	200 in 1892
1865	Asquith	Halifax	200
1865	Dean, Smith, and Grace	Keighley	400
1866	Stirk	Halifax	(Not known but certainly less than 500, the level of the 1930s)
1868	Butler	Halifax	350
1868	Archdale	Birmingham	600
1874	Lang	Johnstone	700
1880	Richards	Manchester	500
1884	Parkinson	Bradford	550
1887	Holroyd	Rochdale	250
1889	Herbert	Coventry	1,600
1890	Ward	Birmingham	650

* The date indicates when the manufacture of machine tools for sale was first
undertaken and does not always, therefore, coincide with the date of foundation
of the firm.

Source: Engineering periodicals and information from the firms. Full details
are available from the author.

buted a major role in the development of the American system
of manufacture to the inventiveness and productive skill of
its machine-tool firms, and the spread of the understanding of
these new methods to other industries through engineers
trained in them. How then did the machine-tool industry in
Britain fare? How was it affected by the differences in the
fortunes of the various branches of the engineering industry,
and how did it, in its turn, determine the nature and rate of
technological innovation? Table 4 gives an outline of the size,

location, and age of the most important specialist machine-tool-making firms in the nineteenth century.

The classic machine-tool builders of the early nineteenth century were all general engineers and within a short time most of them ceased to have much influence on the long-run development of the industry. Clement, Murray, and Fox were short-lived, Maudslay turned to ship-building, Fairbairn concentrated more and more on textile machinery, Sharp Roberts and Nasmyth after 1860 on locomotives. Whitworth's fame in 1851 was enormous, of course, but already his interests were turning elsewhere – to armaments in particular. He had no more to offer the machine-tool industry – indeed he positively subtracted from it. An American engineer touring his works in the late 1860s had this to say: 'Mr Whitworth was not only the most original engineering genius that ever lived. He was also a monumental egotist. His fundamental idea was always prominent, that he had taught the world not only all it knew mechanically, but all it ever could know. His fury against tool builders who improved on his plans was most ludicrous. He must not be departed from even in a single line. No-one in his works dared to think.'[1] The observer went on to write of the poor quality of the machinery used:

> The workmen became indifferent. 'Why do you not renew these worn out bushings?' I asked, but could never get an answer to the question. Some power evidently forbade it, and the fact is that no man about the place dared to think of such a thing as intimating to Mr Whitworth that one of his lathes was anything short of perfect. He had designed it as a perfect thing; ergo it was perfect and no man dared say otherwise.[2]

Through loss of its pioneer firms and the conservatism of its mid-century high priest, the industry lost some of its impetus; in an important institutional sense it had to start all over again. Of course, there were links with the past through apprenticeship and the like. Outstanding among those was William Muir who was with Maudslay, Holtzapffel, foreman to Bramah, and

[1] Charles T. Porter, *Engineering Reminiscences* (New York, 1908), p. 124.
[2] C. Porter, op. cit., p. 129.

works manager to Whitworth, where he was associated with some of the most brilliant developments. Eventually setting up on his own, by 1852 he was supplying tools for interchangeable rifle sight manufacture by the Woolwich Arsenal.[1] New centres of the industry grew up in Yorkshire, created by men trained in the classic machine shops in Leeds, many of them receiving a particular boost in the 1870s when German firms bought heavily of capital goods in Britain. But the new firms were small, the creation of individuals of small resources, though for the most part they now confined themselves largely to machine tools and were not general engineers. An exception here was Greenwood and Batley, founded by men who had been partners with Peter Fairbairn, and possessing sufficient resources to develop their new company quickly.[2] Significantly enough, though, they created a general engineering business in which machine tools accounted for well under a half of their turnover. By the turn of the century they were employing 2,000 men, making them one of the largest of this kind of enterprise in the country.[3]

The industry had surrendered some of the advantages of its early start: how did it now fare? Again, market forces can be seen to have been of major significance. With the exception of the textile machinery firms, who in any case appear to have made many of their own machine tools, the industries already discussed – locomotive, steam-engine, agricultural machinery makers, and to this list one should add ship-builders – brought a demand for the highest quality and most up-to-date machine tools of the heavier kind. Many of the works, as we have seen, were superbly equipped. They generated a machine-tool industry second to none. Buckton, Muir, Hulse, Richards, Asquith, in England, Shanks and Lang in Scotland, were possibly the most famous, and the most outstanding of all was Cravens, who became probably the finest firm in the world supplying heavy machine tools. To give but one example of Britain's superiority, although the American industry pioneered the use of milling for light work, in Britain its use, and the

---

[1] R. Smiles, *The late William Muir, Mechanical Engineer of London and Manchester* (*c.* 1890).

[2] *Mechanical Engineers*, 1874, p. 20.

[3] *Engineering*, 7 August 1903.

production of such machines, for heavy work was by 1900 generally acknowledged to be very much more advanced.[1] Orders for heavy machinery were too small to allow a firm normally to specialize in one type, but even here the British industry was in the van of developments. After 1880 Langs of Johnstone began to concentrate solely on lathes, and were described in a report of 1908 as the only European firm specializing in the American fashion: they were, however, rather special, being also the first company in Europe to supply machine tools with cut rather than cast gears.[2]

But in medium engineering, where the Americans made most spectacular progress, the home machine-tool industry got little support. Although the interchangeable techniques for gunmaking originated in the U.S.A., they were quickly adopted here, first in the government factory at Enfield, and within four years by two private firms – the Birmingham Small Arms Company founded in 1861 and the London Small Arms Company who were already manufacturing interchangeably rather earlier than this.[3] Professor Habakkuk has pointed out that the arms factories here did not become centres of learning as those in America did, but one does not need a deep knowledge of Western history to realize that the Colt Armoury was not just turning out military rifles. The American market advantage was considerable. In Britain the government doled out declining orders between three or four firms to an agreed proportion, and all market and technological initiative was lost.[4] In office machinery little progress was made until American firms set up branch plants in Britain. The watch industry, still entirely a handicraft trade, came under most severe

---

[1] W. F. Durfee, 'The History and Modern Development of the art of Interchangeable Construction in Mechanism', *Transactions of the American Society of Mechanical Engineers*, XIV (1892–3), 1236.

[2] U.S. Bureau of Manufactures, op. cit., p. 213; *The Engineer*, 3 January 1936.

[3] *Engineering*, 26 May 1893.

[4] See generally *History of B.S.A. 1861–1900*. Typescript selections from the Minutes of Directors' Meetings in the possession of the company. From 1864 to 1878 Enfield received three-quarters of all government orders for rifles and the private firms one-quarter. David Landes's criticism of B.S.A. for continuing to do most of its work by hand methods is mistaken. The company history makes it clear that this was the case only in the interim period of waiting until the new machinery was delivered and installed. See D. S. Landes, 'Technological Change and Development in Western Europe, 1750–1914', in *Cambridge Economic History of Europe* (Cambridge, 1965), Vol. VI, 535.

pressure from American and Swiss makers during the 1860s.[1] In 1888 the Lancashire Watch Company was set up to buy out most of the hand workers at Prescott, and began manufacturing by the interchangeable principle. It was soon employing 500 men and the works were said to be entirely equipped with American machinery. Even so, in 1902 some 225,000 watches were made in Britain compared with 2¾ million in the U.S.A. and 6 million in Central Europe.[2] In the lighter sector of the agricultural machinery industry, Britain was left far behind, and the lag in electrical engineering was a serious disadvantage too. All these were the industries which in the United States became the main outlets for the makers of the new machine tools for mass production.

But even where interchangeable manufacture of what we now call consumer durables was successfully carried out, the impact on the machine-tool makers was sometimes negligible. The history of the Singer company is particularly interesting in this respect and little known.[3] The company, founded in the United States, began assembling parts in Scotland in a small way in 1867, and switched to actual manufacture of sewing machines three years later. By 1885 they were making 8,000 machines a week, rather more than the parent factory at Elizabethport, New York. At the turn of the century 7,000 employees were producing 13,000 a week, and the Clydebank factory was by far the largest sewing machine works in the world. Whatever were the original motives for the move to Scotland – the President of American Singer was an emigrant Scot – the existence of relatively cheap labour did not inhibit the use of the most modern machinery. Contemporaries de-

[1] *The Horological Journal* ran articles on the state of the trade during 1867. Two comments are of particular interest. The first (p. 86) pointed to an establishment well known for the use of automatic machinery – 'anyone who doubts the capabilities of machinery to accomplish 60 per cent of the work in a clock or watch would do well to pay a visit to Mr Gillott's pen manufactory at Birmingham.' The second was to the effect that our system against the American 'is as bows and arrows against the needle gun . . . Those who wish the watch trade to exist must promote at once the introduction of capital and an improved system of tools into England: cheap labour is a rotten reed' (p. 103).

[2] *The Lancashire Watch Company Ltd. Its Rise and Progress* (Prescott, 1893), pp. 17–27; *Engineering*, 7 July 1893; *U.S. Monthly Consular Reports*, no. 308, May 1906, p. 90.

[3] Most of the information in this paragraph has been obtained from company records.

scribed the factory with open-mouthed enthusiasm: 'probably the finest monument of a sound invention properly developed that the present age has produced', said one.[1] There was no other factory like it in Britain in the 1880s. The huge number of modern machines employed is shown in Table 5.

TABLE 5 *Singer Sewing Machine Co., Clydebank – Installation of Machine Tools, 1870–1914*

	Milling	Grinding	Drilling	Automatics	
1870–9	216	—	132	90	
1880–9	451	35	245	221	
1890–9	399	428	174	140	
1900–9	710	440	418	441	
1910–14	457	336	248	313	
Total number of machine tools installed to 31 December 1914:					12,390
Total built by Singer at Clydebank:					9,782

Milling Machines:		
Total installed	2,233	
Built by Singer, Clydebank	1,648	
Built by Singer, N.Y.	303	
Built by British makers	179	
Supplied by U.S. makers	38	
Supplied by British agents	63	(Some British made but mostly American)

Source: Company Machine Books.

The early use of grinding machines is particularly worthy of note. But, as the statistics show, well over three-quarters of the machine tools were made by Singers themselves, and a very high proportion of those bought out consisted of straight-forward lathes. The detailed accounts of the purchases of milling machines show that very few indeed were supplied from outside the Singer organization before 1900. The firm therefore had a negligible impact on machine-tool demand; it did nothing towards the building up of capacity and know-how in modern machinery methods in the British machine-tool industry, but simply began to buy once that capacity had been created – largely by the cycle industry.

This was an extreme, but by no means unique, example. Another firm making sewing machines by interchangeable methods was Bradbury of Oldham. Founded in 1852, they were

[1] *The Engineer*, 28 August 1885.

not comparable in size to Singers: in the early 1890s they employed 600 people and turned out 500 machines a week.[1] They too used special machinery designed by themselves and sold machine tools to the trade as a sideline. The tradition of the general workshop making its own tools died hard. The textile machinery firms certainly made many of their own tools. The same can be said of locomotive makers such as Nasmyth, Sharp Stewart, and Beyer Peacock. Tangyes, as we have seen, began their machine-tool business in this way. More famous still was the Birmingham firm of Nettlefold and Chamberlain which utterly transformed the wood-screw industry, using techniques acquired by buying in 1854 patent rights to American automatic machinery. As early as 1869 they were employing milling cutters to shape all six sides of nuts simultaneously, but a report of 1876 significantly noted that 'all the machines and tools are made at the works'.[2] For all these reasons the commercial production of new medium machine tools – milling machines, turret lathes, and later, grinding machines – got off the ground only slowly. Their value was appreciated by the best shops: importing agents such as Churchills did much to make them known. Many of the new machine-tool firms took up their manufacture. A report on the Vienna exhibition of 1873 noted that 'the workshops of certain leading firms in England are being filled up with tools of a special kind, possessing great originality with regard to fitness for a purpose in the manufacture of general machinery, agricultural engines, small arms, etc.'[3] But the break-through of mass demand was slow to come. The machine-tool firms remained small in size and limited in resources. Even makers of outstanding brilliance such as Smith and Coventry, very highly praised by the American, Charles Porter, and by Alfred Herbert, the firm which introduced twist drills to Britain in 1876 and did more than any other to popularize milling, even they found it impossible to specialize.[4] Long runs were essential to cover the cost of the special machine tools, the elaborate jigs and templates, and the time taken to set them up – and this applied as much to the

[1] *Oldham of To-day* (Oldham, 1898), pp. 21–3; *The Ironmonger*, p. 54.

[2] *Mechanical Engineers*, 1876, p. 328.

[3] *Reports on the Vienna Universal Exhibition of 1873*, pt II, P.P. 1875, LXXIII, 12.

[4] C. Porter, op. cit., pp. 166–76; *Machine Tool Review*, XLIII, September–October 1955.

makers as to the users of machine tools. It was, too, a vicious circle, for these small firms were inadequate training grounds for engineers to go out and spread the gospel of the new production methods.

The break-through in demand came from the cycle industry. The arrival of the safety bicycle and the pneumatic tyre created such a boom that big makers in Coventry and Nottingham were forced to reorganize their methods completely to meet the demand. Small men began to assemble and specialist component makers grew up to supply them. A firm such as the Birmingham Small Arms Company, which had trifled with cycle-making in the 1880s and abandoned it altogether in 1888, came back into the trade. By 1896 they were making 2,000 sets of cycle components a week and buying in large quantities of machine tools – giving even a small firm like Holroyd a single order for 124 milling machines in 1896. Vast capacity was created by the large cycle makers and rows of the best machinery installed as a result of the heavy investment of the early 1890s. The financial bubble burst, but now the pressure was on: the need was to use this machinery to the full, to standardize and to extend the market, and this they did remarkably well. In 1913 Britain exported 150,000 cycles, Germany 89,000, and the rest of the world almost none at all. Much of the early machinery was imported, but quickly new firms such as Ward's and Herbert's arose in the Midlands and older firms flourished under the new impetus. Interchangeable batch production and production for stock for the first time became standard machine-tool-making practice. Nor was it just the demand of cycle-makers. The belated development of electrical engineering, the exacting requirements of steam-turbine and gas- and oil-engine builders all helped. Then, too, came the motor-car industry. One must be careful not to fall into the error of underestimating this development simply on account of the more spectacular progress made in the U.S.A. Table 1 shows that by 1907 motor-car manufacture was already a sizeable part of the engineering industry and in the next six years output was to increase almost threefold. Nor was the running all made in the medium sector of the industry. Many makers of heavy machinery, too, rose brilliantly to the market opportunities offered by the development of high-speed steel, taking advantage of their reputation

for strength and rigidity, and redesigning their tools to trans-
mit the greater power now required. Tangyes, with huge high-
speed lathes, Hetheringtons with high-speed radial drills, and
many Halifax firms were to the fore here.[1]

In this way the machine-tool industry reached its maturity.
Of course, not every problem was solved, not all the leeway was
made up. Specialization remained rare: except for Herberts and
Cravens, the average size of firm was low. As Mr Rosenberg
has pointed out, the degree of specialization and of size
achieved in the United States was possible only because of the
simultaneous growth of several industries enjoying in common
certain technical processes.[2] The American advantage, once
established, was difficult to break. In specialized lines, world
demand can easily be satisfied by one firm enjoying considerable economies of scale – indeed, much of world trade in
machine tools has been, and still is, of this kind. Simply for
this reason, unlike many other new engineering industries,
no American machine-tool firm found it worthwhile establishing a branch plant here, but concentrated on arranging agencies
for sale of their special-purpose machines. It is significant that
Herberts, the largest manufacturers, were before 1914, and
still are, major agents too.

The development of the machine-tool industry after 1850
can be seen, therefore, to reflect partly a discontinuity in growth
and partly the pattern of demand for heavy and medium
machines. Of course, demand is not just something objective
faced by a manufacturer, but something that he can, at least to
some degree, seek to fashion to his own liking. One may blame
the machine-tool makers for not doing enough to take up,
develop, and force new ideas on their customers. No doubt
there is some truth in this line of argument, but could the break-
through really come in this gradual way? Must it not force its
way through certain leading sectors such as arms were in the
U.S.A.? In Britain it might have been sewing machines, but
in fact it was cycles.

---

[1] *The Engineer*, supp. November 1905, p. i; *American Machinist*, 1 February
1906, p. 163.

[2] N. Rosenberg, 'Technological Change in the Machine Tool Industry, 1840–
1910', *Journal of Economic History*, XXIII (1963), 424.

III

We can hardly argue in terms of the market without at least a passing reference to selling techniques, though there is no space here for a detailed analysis. Even a superficial examination of the sources, however, indicates that the usual complaints about British methods are too general to be convincing. Some writers argue that not enough was done to determine and satisfy the needs of particular markets: others bemoan the lack of standardization. We cannot be too dogmatic on this: for agricultural machinery, differences in physical environment required considerable product variation and British firms apparently responded well. In much textile machinery, such was the reputation of Lancashire and so unimportant the physical environment that standardized equipment was the rule, though of course this could only be created by extensive sales efforts. A firm like Mather and Platt had its resident staff of engineers in India: senior members of the firm travelled extensively in Europe and South America: In Russia their agent was Ludwig Knoop. You could hardly better that. Yet this did not apply to loom-making, most of which was carried out on a small scale and quite unstandardized largely because local specialization in weaving had encouraged local loom-makers to cater for their own districts, and many different types gradually emerged – a pattern only broken when an entirely new form of loom appeared such as the Northrop, which was standardized from the first.[1]

By 1880 the specialized selling agent was becoming more and more the rule, and for the largest firms, such as Tangyes and Marshalls, direct representation overseas was becoming extensive. Many examples could be quoted of quite small firms – 300 employees – with showrooms in London and elsewhere in Britain and many agents overseas.[2] It is sometimes said in criticism that the agent had no particular incentive to push a firm's goods over those of its competitors and was a barrier to technological change. The point is not convincing: firms could always change agents, and many did. Garretts,

[1] The 1915 catalogue of Geo. Hattersley and Sons Ltd of Keighley showed 34 different kinds of loom not including minor variations.

[2] *Victoria County History of Warwick*, Vol. VII (1964), 190.

agricultural machinery makers, made one particularly strong argument in favour of agents: orders came from them in blocks and enabled the makers to plan a production programme: in the boiler shop, flanging blocks could be set up for a particular type and as many as 100 boilers made in one run.[1] The old jibe about catalogues in English does not always ring true either. Mrs Smith has shown this was not so of many Birmingham firms. As early as 1853 Ransomes had catalogues in French, and by 1873 in Russian too: in 1862 Garretts had catalogues in French, German, Danish, and Spanish – agents in Paris, Sydney, Calcutta, New Brunswick, Taganrog, and an office in Pesth, and the other agricultural machinery firms were doing the same.[2] As for the cycle and motor industries, they indulged extensively in trials, races, exhibitions, clubs, and built up hire-purchase facilities. The cycle industry in particular had a wide range of sales literature masquerading as journals – *Cyclist, Cycling, Bicycling News, Wheeling, Cycle Record, Wheel World, Northern Wheeler*, and so forth.

Of course one faces a serious problem of contradictory patterns of behaviour. There were the textile machinery makers imposing their standardized equipment on buyers, but making it in a conservative manner: the locomotive builders more advanced in an engineering sense, but having to conform to customer preferences. Some of this has been explained by way of institutional and market patterns, but we are still left with the problem of distinguishing between the best and the average, of discovering how representative is the pattern established in this article. One gets the feeling that the general engineering jobbing shop was a peculiarly persistent feature of British engineering and certainly no place to sell the latest in machine tools. Alfred Herbert described one where he was an apprentice:

Jessops employed about 150 men. They turned out really good work by the standards of their time, but by comparison with a modern shop their equipment and their methods were primitive. There were no milling machines, no capstan or turret lathes, no grinding machinery, no gear-cutting

[1] R. A. Whitehead, op. cit., p. 90.
[2] *Victoria County History*, loc. cit.; R. A. Whitehead, op. cit. p. 74; *Ransomes' 'Royal' Records*, p. 86.

machines, for all gearing was cast, no twist drills, no jigs or fixtures, and not even a blue print. I worked on an 8-inch sliding and surfacing lathe – a really good tool, but like most of our machines sadly in need of reconditioning. The bed was badly worn and to turn a parallel job the calipers were constantly in use, then the work was filed while rotating and finally polished with emergy cloth or by clams, a pair of big wooden nut crackers fed with a horrible mess of emery powder and oil.[1]

Here is another description:

In the ordinary machine shop of that day there were, generally speaking, only three types of machine tools – engine lathes, drilling machines, and planers. Milling machines as well as shapers were scarce, and automatic screw machines were looked upon with wonder . . . Turret lathes as we know them today were unknown . . . Even in reasonably well-equipped machine shops of that day twist drills were not in use . . . All measuring was done with outside and inside calipers and scales, the micrometer being unknown in the ordinary shop.[2]

The difference is that this is what James Hartness described as the average American machine shop about 1880. If this article has concentrated on the best in British practice, it is not clear that writers on German and American industry have not done the same. Enterprises such as those just described had an obvious niche in the market: the quality of work was often high, and with no real backlog of orders they could carry out single orders quickly. Whether they were operating as efficiently as they might, even allowing for the limitations of the market, is an open question. It may well be that they fitted in better and longer to the less specialized pattern of British engineering with its plethora of one-off orders, and that therefore their conservative influence was the more powerful. Certainly it became a self-perpetuating pattern of activity. Buyers got into the way of insisting on special features and it was a habit which the young, small machine-tool firms, living on a precarious margin were in no position to break.

[1] *Machine Tool Review*, XLI, September–October 1953.
[2] Wayne G. Broehl, Jr, *Precision Valley* (Englewood Cliffs, 1959), p. 12.

G

IV

My aim has been to depart from vague generalizations about British industrial performance and to point out differences between the various sectors. I have tried to show that some sectors of British engineering – makers of textile, steam, and sewing machinery, for example – were very advanced commercially and technologically, and that where in others – agricultural machinery, locomotives, and above all in that most critical sector, machine tools – the degree of success varied, it was in part at least conditioned by the nature of the market. Much remains to be explained – the weaknesses in those mass-production industries I mentioned before, watches and office machinery. The gunmakers may have been at a disadvantage in their own field, but why did it take them so long to apply their techniques elsewhere in the way that Remington did? As for motor cars, I have elsewhere questioned the adequacy of the market as a complete explanation. In other sectors – electrical machinery, for example – the market has been cited as a factor in slow growth, though here too it is obviously not the whole answer in view of the striking successes achieved in the north-east in a more helpful institutional environment. Peter Temin's recent study of the steel industry takes the pattern of demand as a neglected factor, but does not deny the stupidities and failings of the industry itself.[1]

Clearly the market is not a complete answer to all our questions. To understand more fully the lags and successes of British engineering there is need for more detailed study of its training and institutional patterns. The usual and often justified condemnation of the system of formal technical education must be modified so as to distinguish between the needs of different products – the bicycle essentially the product of practical men – the steam-turbine which was inevitably the province of the trained engineer – the early internal combustion engine developed by professional engineers but improved in major aspects by practical men. The role of the consulting engineer has already been briefly mentioned, but there is abundant evidence of his baleful influence in other sectors

[1] *Industrialization in Two Systems*, ed. H. Rosovsky (New York, 1966), pp. 140–55.

too – civil and electrical engineering, for example. The traditions of Scottish heavy engineering created by locomotive and ship-building were impervious to the lessons the Singer factory offered: to Singer's considerable misfortune, the opposite was not the case. Most engineers to 1900 still came up through an apprenticeship, and the machine-tool firms developing the American methods after 1870 were too small to be an effective training ground, and this is the one outstanding contrast with American experience. Importing agents such as Churchill's and Buck and Hickman played an important role in spreading knowledge of the new techniques, but it was not the same thing. You do not learn from talking to an agent as you do from working with a maker. In Britain the railway workshops and the big steam-engine shops were typically the places where young men served apprenticeships. It is not without significance that Stoke City Football Club was founded by a group of public schoolboys serving their time in the local railway workshops: Royce, Bentley, Austin, and A. V. Roe of aircraft fame did the same; Singer, Hillman, the Starleys, Willans, were among the many apprenticed at Penns, marine engineers of Greenwich. These were first-generation men, first-class engineers, men of drive and enthusiasm, but trained in a traditional environment and finding it hard to break with the past.[1] The American motor-car industry, on the other hand was fertilized not by college graduates but by engineers trained

---

[1] This is not to argue by any means that no direct transference of interchangeable skills took place. B.S.A. and Bradbury's we have mentioned. As one report said of the latter, 'For over 30 years the firm have turned out their famous sewing machines in the system of interchangeable parts and they have now applied the same principle to the construction of cycles' (*Oldham of To-day*, p. 23). They also turned out automatic and capstan lathes, drilling and profiling machines for cycle-makers. Harry Lea of the Lea and Francis cycle company was formerly employed by Singers, and Graham Francis had worked for both Pratt and Whitney and Ludwig Loewe (*Machine Tool Review*, XLII, January–February 1954). More interesting still was the career of George Accles, apprenticed at Colt's works in Hartford and engaged later in establishing works for manufacture of cartridges and of the Gatling Gun in many parts of the world. Eventually he set up his own works in Birmingham in 1888 but soon turned to the manufacture of cycles and of labour-saving American-type tools for the industry. His works were described with much enthusiasm by Alfred Herbert (information from Mrs Barbara Smith and from articles in the *Warley News*, 8, 15, 22 April and 1 July 1965). It is not without significance that there were no major railway works in Birmingham and the Black country to impress their traditions of engineering there.

in modern machine-tool shops and consequently capable of appreciating the complete reorganization of the production processes that the new machine tools required. Here, it was slow to come. The leading firms in the British motor industry, controlled by these men of high ability, their shops packed with rows of the latest machinery, still operated without comprehension of what modern engineering implied. Love of the technical product rather than the technique of production was slow to disappear.

I have not tried here to present a general apologia for British industry. This article has sought to throw some light on areas of British engineering that should be better known. If it has played up the successes, technologically and commercially, it is because they are generally underestimated. If the emphasis has been shifted from entrepreneurial explanations of Britain's relative decline, it is to readjust the emphasis and seek more precision, rather than in any way to cast it aside completely. It may well be, as Professor Postan argued in a paper presented to the Manchester Conference of 1966, that in all explanations of relative responses to technological change, the firm and the individuals in the firm remain the residual factor. Unfortunately this does not help us to determine how important that residual factor is. This article has tried to analyse in some instances the environment in which the firm operated, seeking to define more closely the extent to which the residual factor of enterprise must be called upon for explanation. It may be that mechanical engineering differs in important respects from other industries, but in that field at least I would argue that the slowing down of British industrial growth after 1870 was due more to objective economic factors than has previously been recognized.

Colt's London Armoury

HOWARD L. BLACKMORE

[This article was first published in the *Gun Digest* 1958.]

By the middle of the nineteenth century, Samuel Colt was firmly established as a manufacturer of revolving firearms in his own country. His factory at Hartford was running smoothly under his new superintendent Elisha Root, and it seemed a favourable moment to consider the conquest of new markets in Europe and the East. As early as 1835, he had taken out patents in London and Paris, and in 1849 he arranged similar protection for his new model revolvers and rifles.[1] The success of the U.S. Ordnance trials of his revolvers was known in Europe but the various import restrictions and duties had prevented any substantial sales, so that the guns themselves were comparatively unknown. Remedying this perplexed Colt for some time.

In 1851, however, a giant iron and glass building known as the Crystal Palace was erected in Hyde Park, London, to house a Great Exhibition of arts and crafts from all over the world, and Colt's problem was solved. The leading gunmakers of Europe sent examples of their work and were allowed a duty free stay while on exhibition. It was expected that military and naval representatives from all the European countries would be visitors, and Colt determined to make the most of the opportunity. In the United States of America section, therefore, among such exhibits as Dennington's Floating Church for Seamen, Goodyear's India Rubber Life Boat, and Mr Perkin's Steam Gun, pride of place was given to his 'New Repeating Firearms'. Unlike his European rivals, who sent one or two magnificent specimens, Colt brought over several hundred guns. Some were presentation pieces, plated and engraved, but the most were plain production models, arranged impressively around the walls of his stand. The purpose of this panoply

[1] British Pats. No. 6909/1835, No. 12668/1849.

of arms was twofold. First, to demonstrate the mechanical qualities of the revolvers and the interchangeability of their parts; secondly, to emphasize the fact that they could be made in quantity by machinery.

This was a sensational idea to Englishmen, whose military and sporting firearms were traditionally hand-made, involving the employment of dozens of different outworkers. Since the days of Marlborough, English armies had been supplied with weapons by the Board of Ordnance, which gave orders to contractors in London and Birmingham. As a general rule, the barrels, locks, and miscellaneous metal parts were made in Birmingham, then sent to London where they were rough stocked and set up. Whenever a new war caused a rush of orders, the whole system broke down and the Board was forced to buy arms from one of the Belgian or German arms-making centres. Interest was aroused in Colt's revolver, therefore, not only because of its reported performance but by the suggestion that several hundred could be made in one factory in a single week. The English gunmakers, of course, laughed this to scorn; but not for too long a time!

The Great Exhibition opened in May 1851, and official circles were not slow in taking notice. On 5 June the Admiralty asked the Board of Ordnance to procure some of 'Mr Colt's repeating pistols' for trial.[1] Eight revolvers with flasks, at £6.10s.0d. each, were purchased from Charles W. Lancaster, the London gunmaker, apparently Colt's agent at this period.[2] Four of these were sent on 8 July to Admiral Sir T. B. Capel at Portsmouth, to be tested by Capt. Chad on board H.M.S. *Excellent*.[3] The other four were presumably tried out by the Ordnance officers but no record is known of either trial.

At the same time, the sporting gentry led by Lord Ranelagh organized a series of trials at Mulgrave House, Fulham. Here Sear's needle gun and a French pin fire gun (probably Lefaucheux's) were put through their paces and a certain Mr Deane produced a revolving pistol which put four shots out

---

[1] Public Record Office, London. Minute Books of the Board of Ordnance, W.O. 47/2259, p. 5767.

[2] W.O. 47/2261, 30.6.1851, p. 6563.

[3] P.R.O. Admiralty Letter Book, Military Branch, (Home Stations), Adm 2/1558.

of five into a three-foot target at 100 yards.[1] *The Times,* in its
report of these trials, made a pointed reference to the fact that
the Colt revolver was not displayed and somewhat disparagingly
remarked: 'It was stated by some person on the ground that so
far from being an American invention there is an English-
man living who devised the plan thirty years ago.'[2] It was the
beginning of a rivalry between Colt and the English revolver-
makers which was to continue bitterly for several years.

But Colt was not unduly worried – he had his eye on the
main chance. At the end of June, the Board of Ordnance
arranged for 25 of his revolvers to be released to Lancaster,
who in turn sold them to officers of the 12th Lancers, under
orders for South Africa.[3] This was followed by many sales to
individuals, authorized by the Treasury, who gave Customs
permission to let the revolvers leave the Exhibition duty free,
providing the buyer was an officer proceeding overseas.
While this was good publicity for Colt, his main objective
was to win the approval of the Select Committee on Small
Arms at Woolwich, who tested all new weapons for the
Board. To this end he persuaded the American Minister in
London, Mr Abbot Lawrence, to give him some help.

On 1 September, 1851, the Master General of the Ordnance, Lord Anglesey, sent the following letter to Colonel
Chalmer, the secretary of the committee.

My Dear Col.
    The American Minister Lawrence has just been with me
& is most anxious that I should see & hear a Mr Colt, who
is the Inventor of the curious American Pistol.
    I am going out of Town to-morrow. Mr Colt has various
letters of introduction to me. I explained to Mr Lawrence
that all inventions are laid before our Select Committee of
which you were the Secretary & I have agreed with him
(Mr L.) that you should open all the letters Mr Colt has to
present to me & that you should act upon them. You will
find Mr Lawrence a very fair and intelligent agreeable
Person and I recommend all these matters to your good
offices giving me an account of what may pass regarding these

---

[1] *The Artisan Journal,* London, 1852, No. VIII, Vol. X, p. 169.
[2] 11 and 19 June 1851.
[3] *The Times,* 27 June 1851.

small Pistols. Mr Colt it seems has got a Patent & he is desirous of being allowed (if the weapon is approved) of making in this country what may under his own supervision be wanted for its Service

I remain etc.[1]

The London gunmaker Robert Adams had also submitted a revolver and when the trial began at Woolwich on 10 September, it became not so much a trial of Colt's revolver but of Colt *v.* Adams. According to *The Times* report of 11 September, the following persons were in attendance: Lt.-Gen. Sir Thomas Downman; Major-Gen. Fox; Col. Dundas; Lt.-Col. Chalmer, Asst. Dir. Gen. of Artillery; Lt.-Col. Burn; Capt. Wingfield; Capt. Anderson R.A.; Brig.-Major Walpole R.E.; Capt. Fox, Gren. Guards; Capt. March; George Lovell, Insp. of Small Arms; and Mr Lawrence, Junr., the son of the American Minister.

Colt commenced, firing his own revolver at a target 6 feet square from a range of 50 yards. The Adams revolver was then demonstrated by a Mr Winter. When the trial was completed and the party was leaving the butts, an incident typical of Colt occurred. A company of the Royal Sappers and Miners, proceeding on embarkation, was encountered, when 'Mr Colt in a very handsome manner, with the consent of Major-General Fox, presented Lieutenant Ray, in charge of the company, with one of his revolving pistols.' Unfortunately, *The Times* fails to give the trial results, and the only account appears to be that published by Deane, Adams & Deane, makers of the Adams revolvers. According to this, their revolver showed such a marked superiority that one must assume Mr Winter was a better marksman than Colt or that the report was exaggerated a bit.

Either way, damage was done to Colt's reputation, as is revealed by the letter sent by the Master General's Secretary to Sir Charles Wood on the 16th of October 1851:

Dear Sir Charles

I have seen the Secretary of the Small Arms Committee & he reports that the *present* Revolvers are not calculated for our Service on account of the Smallness of the Nipple

[1] P.R.O. Out Letter Book of the Master General of the Ordnance, W.O. 46/86.

& Cap but that the experiments are not brought to a conclusion & that the Committee are not prepared with a final
report on the subject.

I have therefore informed Col. Sam Colt that the Ordnance are not prepared to purchase any of the Arms that are
at the Exhibition.

I may privately inform you that there is another revolver
before the Committee which bids fair to beat the Colt . . . [1]

Officially rebuffed, Colt determined to impress the merits of
his revolver on the various scientific and military societies in
London, and he did not hesitate to resort to a little gentle
bribery. In a letter to Sir Charles Trevelyan, Secretary to the
Treasury, on 15 October 1851, he applied for permission to
withdraw the following articles from the Exhibition:

. . . one Rifle and two Carbines intended to be used as
Models for manufacturing similar Arms in England which
I have no desire to sell.

Fifty Pistols – assorted sizes, engraved and highly finished,
which it is my wish to present to various distinguished
gentlemen in England – to Institutions such as the Institute of
Civil Engineers, the United Service Club Museum etc. etc. –
and to some of the Commissioners of the Foreign Departments at the Exhibition for their respective Governments.

About 250 Pistols in assorted sizes – in my Show Cases
in the Exhibition for the purchase of which I have received
application from many Officers of the Army and Navy and
other distinguished gentlemen in England whom I should
like to gratify – and whom I have promised to notify if
permitted to dispose of them.[2]

It was to the Institution of Civil Engineers that he turned his
main attention. Their secretary, Charles Manby, became his
friend and agent. His first job was to obtain permission from
the Inspector General of Fortifications on 22 October for Colt
to 'examine minutely certain old arms at the Tower and to use

---

[1] P.R.O. Out Letter Book of the Master General of the Ordnance, W.O. 46/86.
[2] Kings Beam House, London. Minute Book of the Board of Customs, 22.10.
1851. According to some original correspondence in the Ray Riling Colln.,
Samuel Colt signed an affidavit on 28 October to a list of presentation pistols
numbering 86. See Ray Riling, *Guns and Shooting* (New York, 1951), No. 606.

them as illustrations of some paper he is proposing to read to the Society.'[1]

On 25 November, Colt gave his famous lecture to the Civil Engineers. His theme was not only the comparison of his own revolver with those of the past but the advantages of making them by machinery. Among the audience was his rival Robert Adams, who was also given the chance to display his revolver. The latter went further and distributed copies of his own account of the trials at Woolwich. This drew forth a denial from Sir Thomas Hastings, a member of the Board of Ordnance, who was present, that the report was an official one and several members spoke up in support of the Colt revolver. There is no doubt that Colt gained sympathy and respect from this lecture for in the following year, he was elected an Associate of the Institution and awarded its Telford Medal. From some of his remarks during the lecture it is obvious that he still had hopes of persuading the Ordnance to adopt his revolvers 'after being adapted to the standard percussion cap of the service'. He had also gained the approval of Col. Chalmer, who had spent a week firing his revolvers with great success and had obligingly sent copies of the targets to the Institute.[2]

Colt ended the year 1851, therefore, in a happy mood, distributing his revolvers to those whose influence he hoped would help him in the future. These included Lord Anglesey, the Master General, who sent a short note begging 'to express his thanks for the Colonel's obliging attention in sending his specimens of his Pistols'.[3]

In the new year, Colt received permission from the Treasury to export some 450 of his revolvers left in the Exhibition building duty free, to his agent in South Africa, a Mr Peard.[4] From a report in the *British Army Dispatch* of 10 December 1852 this gentleman also had his difficulties:

In appears that Mr Peard, Colonel Colt's agent in Southern Africa, after having met with much difficulty in establishing

[1] W.O. 47/2273, p. 10241.

[2] Col. Sam Colt. 'On the Application of Machinery to the Manufacture of Rotating Chambered-Breech Fire-Arms', *Minutes of Proceedings of the Institution of Civil Engineers*, XI (London, 1853).

[3] W.O. 46/86. 1 January 1852.

[4] Customs Minute Books, 6.1.1852; *The Times*, 14.1.1852.

the reputation of the arm, owing to base and unfair repre-
sentations on behalf of interested parties, at length brought
the matter to public proof or test in a place called Baakers
Valley. There the firing commenced at 200 yards! Out of
one cylinder of six charges, two bullets were put in the
target by Mr. Peard at this distance.

Another newspaper, the *Graham's Town Journal*, 2 October 1852
also reported:

> The penetration of one of the second size pistols, wt.
> 2 lbs. 6 ozs., weight of bullet one third of an ounce was tried
> at the Hoek by Quarter Master Serjeant Rennie and found
> to be greater than that of a musket with a round ball.
> The pistol projected the bullet through $\frac{3}{4}''$ hardwood board,
> behind which were two blankets folded, as carried by
> soldiers, eight folds each – the bullet went clean through.
> The musket ball stopped in the fourth fold of the second
> blanket.[1]

From all accounts, Colt had every reason to believe, once
the initial prejudice against a new weapon had been removed,
that there would be a large market for his revolvers in England
and her colonies, and what better place to make them in than
the very hub of the Empire – London. Colt therefore, established an office and showroom in No. 1 Spring Gardens and
appointed Charles Frederick Dennet, an American living in
London, as his manager. Then he began the business of setting
up a factory. Charles Manby found a group of buildings on the
Thames Bank near Vauxhall Bridge, which seemed ideal for
the purpose. This property, under control of the Office of
Works and Buildings, had been used by workmen during the
building of the new Houses of Parliament. As that work was
nearly finished, part of the premises was being used for storage
purposes only. On 18 December 1851 Manby wrote to the Office
of Works to ask them if they would lease to Colt 'that portion
of Thames Bank Workshops then only occupied by old casts.'[2]

---

[1] Excerpts from an undated 8-page pamphlet headed '*Improvements in Small
Arms, Colts Repeating Pistol*'. It is possible the agent Peard was the same man as
the A. T. Peard listed as a workman at Whitneyville in 1847. (See *Sam Colt's Own
Record 1847*, edit. John E. Parsons, 1949, C. H. S. Hartford).

[2] P.R.O. Misc. Papers, Ministry of Works 6/160/9.

Some correspondence then took place with Sir Charles Barry, the architect, who finally agreed to move the casts elsewhere and clear some of the buildings for Colt to use. As soon as this was known, Manby notified the Office of Works of some alterations which Colt wished to make, notably the installation of a steam engine and a boiler house in the north end of the main building. As some of the buildings to the west were still occupied by carvers and joiners, it was agreed that a dividing wall nine feet high should be built right through the property, leaving the main entrance in Grosvenor Road for the use of the Office of Works, whose Clerk had a house nearby. Early in February, Colt, Manby, and an Office of Works architect named Pennethorne visited the workshops and after deciding to move the proposed position of the dividing wall twelve feet more to the west, agreed on the final plans. Pennethorne recommended that Colt should be granted a lease on terms of which the following is a summary:

1. A rent of £250 per annum plus £5.5.0 for the additional ground (due to the moving of the boundary wall) on a lease of 7, 14, and 21 years.
2. All rates and taxes to be paid.
3. The premises to be kept in repair.
4. The premises to be insured for £4000.
5. No chimney to be built higher than the adjoining one.
6. No opening to be made in the wall next to the public street.
7. No injury by smoke, noise, etc. to be caused to the adjoining property held by Mr Cubitt, etc.[1]

With this settlement in view, Colt returned to America, leaving the final details of the lease for Manby to deal with. Weeks went by with no word from Manby, and by the end of March Colt was near to despair; writing to Dennet from Hartford, he advised him to keep the name of Colt in the public eye with shooting exhibitions and newspaper notices and assured him that 'the instant that I hear that all is right on the otherside I will start off men & Tools to commence the manufacture of my arms in England.'[2]

---

[1] P.R.O. Misc. Papers, Ministry of Works 6/160/9, 6.2.1852.
[2] William B. Edwards, *The Story of Colt's Revolver* (Harrisburg, 1953), p. 297.

In London, however, legal complications had arisen. The Office of Works were themselves tenants of the Commissioners of H. M. Woods, Forests & Land Revenue, and before subletting part of their premises to Colt were anxious to put their own lease in order. Correspondence started between the two Departments which continued until the last years of Colt's tenancy. But on 1 May, T. W. Phillips, Secretary of the Office of Works, was confident enough to write to Manby and confirm the clauses of the proposed lease.[1] Two days later, Manby replied:

> I beg to subscribe entirely to the terms stated and am prepared on behalf of Colonel Colt to enter into the necessary engagements and to fulfill the conditions of the Agreement for a Lease which will of course be prepared in due time if it were possible to expedite the obtaining possession of the premises I should feel much obliged as I feel that much time has been unavoidably lost in the preliminaries.[2]

No objection was raised to the immediate occupation of the premises and Colt decided to go ahead with his plans, lease or no lease. He wrote to Manby to thank him and express the hopes 'that you have treated the Babes of the Woods and Forests with a small specemin of sivelesation in the shape of a Diner at my expense.'[3]

The rest of the year was occupied with the installation of machinery and the training of unskilled labour by the American mechanics whom Colt brought over. The workmen he recruited were butchers, clerks, servants, in fact, anyone but a gunmaker, and they might well have been savages for all their knowledge of machinery. Writing from London on 17 November 1852 Colt fulminated:

> Dam me if I believe it would ever have been made if it fel to the lot for Englishmen to take the Job . . . they say here when the Lion wags his tale all Europe trembles . . . it

---

[1] The lease of the factory cannot be found and this letter of 1 May and Manby's reply constitute the only known agreement between the two parties. Parts of the original of the Office of Works letter, in the collection of the Connecticut Historical Society, was published by Edwards.

[2] Works 6/160/9. 3.5.1852.

[3] Edwards, p. 296.

is my humble opinion that he may wag and be damed so far
as mecanick arts are conserned his biggest flerish would not
scare the youngest of the Yankey Boyes I brought here with
me & the old "critter" will soon begin to understand she is
yit to receive instruction & Pap from her oldest child in her
second childhood.[1]

His letters, however illiterate, were always illuminating.

However, he persevered; his 'men of ignorance' began to
acquire the necessary skill and on 1 January 1853,[2] factory production started. The whole project aroused the curiosity of the
public and the press. Charles Dickens may have been conducted round the factory, for he describes the brick barrack-like
building and the 30 h.p. steam engine, 'indefatigably toiling
in the hot suffocating smell of rank oil down in the little stone
chamber'. This conveyed power to the long vista of machines,
'treating cold iron everywhere as if it was soft wood, to be cut
to any shape without straining a muscle'.[3]

From this report and others, a description of the factory can
be built up. The main building, which ran alongside Bess-
borough Place, consisted of three floors and a basement. At the
north end, stairs gave access to all floors and under these the
steam engine had been built, receiving power from the Boiler
House which Colt had erected on the vacant space near the
gateway. In the basement were planing machines and heavy
lathes used in making new tools, etc. On the first and second
floors were the turning, boring, and milling machines, as well
as five rifling machines each capable of rifling 100 barrels a
day. On the top floor the polishing and final assembly was done.

Along the dividing wall, Colt built sheds for the five steam
hammers and a drop forge used in making the actions and

---

[1] Edwards, p. 301.

[2] This is the date given by Colt in an answer to the Parliamentary Committee
on 16 March 1854 (See p. 181, note 3). It is doubtful, however, that the factory
was in full operation by this date. To the same Committee, Colt gave evasive
replies to questions concerning his first year's production. On 1 January 1854,
an advertising leaflet from the London Factory offered for sale five different
types of revolvers – the Navy and Cavalry models with 7½-inch barrels and
three pocket models with 6-in, 5-in and 4-in barrels. A special note on the leaflet states that the last three would not be ready until 1 February 1854.

[3] *Household Words*, London, conducted by Charles Dickens. Week 27.5.1854,
pp. 353–6.

barrels. At the south end of the main building, at right angles to it, was a smaller block, used mainly for offices and show-rooms. This was separated from the main road by an ornamental garden. On the other side of the road was a series of wharfs lining the river bank. One of these was offered to Colt but he refused it. An interesting feature of the factory was the proving room where the pistols underwent a preparatory trial before being sent for the regular government proof. A later map of the factory shows a proof house in the south end of the basement area, situated on the west side of the main building.[1]

The factory was conducted on lines which would compare favourably with those of today. The various parts moving through the factory were subject to rigid inspection. The factory hands were given the same strict surveillance and an ingenious clocking-in timepiece kept check on the supervisors and watchmen. To offset the rigours of the ten-hour day, a reading room and baths were provided.

With the factory properly organized and a production of 1,000 revolvers a week possible, Colt badly needed a large contract. Unfortunately, in 1853, the Board of Ordnance was engaged mainly in deciding which rifle to adopt. With 'a cautious vigilance which so important a national question deserves' it was carefully experimenting with different forms of rifling and cartridges, and was in no hurry to adopt the revolver.[2] It was also considering the building of a Government factory to make arms by machinery and do away with the old contract system. A commission was sent to the U.S.A. to inspect the methods of making arms there and also to order the necessary machinery.

All this was not to the liking of the established gun trade. With the aid of their M.P.s the gunmakers succeeded in calling an inquiry into the business. A Parliamentary Committee on Small Arms was set up to investigate the whole question of arms manufacture in England.[3] Colt appeared as a witness and his London factory was cited as an example of what could be

[1] P.R.O. Supply 5/60. Map No. 7, 1865.
[2] W.O. 46/89, 2.8.1852. Viscount Hardinge to Col. Sykes.
[3] See the *Report from the Select Committee on Small Arms*, Printed for the House of Commons (London, 1854).

done. The other gunmakers questioned did their best to discountenance him. But as the patent agent A. V. Newton pointed out, Colt's aim was 'not merely to make all the like pieces counterparts of each other but also to banish, as far as may be, the file from the workshop and thus while greatly expediting and reducing the cost of manufacture to render himself perfectly free of all combinations of skilled workmen.'[1] It was an endeavour wholeheartedly supported by the Ordnance Office, harassed once too often by the failure of the gunmakers to complete their contracts for Minié rifles in time. However, much as they approved of Colt's methods, it did not follow that they were prepared to buy the guns he was making. The task of persuading them to do so now became for Colt the all important issue. How this was eventually achieved was described in the *Scientific American*, 16 January 1864. An unknown correspondent wrote:

> One evening at a place of public entertainment I was introduced by a brother member of the press to Colonel Sam Colt who had recently established a manufactory of arms on the banks of the Thames. The revolver had achieved much commendation from the small fry of the press, the starving weeklies which would praise anything for the sake of an advertisement; but Colt was sagacious enough to know that one line in the *Times* would do him more good than a column in any other paper. But how to get to the *Times* except in the advertising columns was the difficulty. Colt was ready to pay any sum that might be demanded for a paragraph . . . As I happened to be well acquainted with some of the *Times* staff and also with two or three M.P.s who had the ear of John Delane, the Editor, I told Colt that I would make an effort to get him fairly noticed . . . (I) was turning over in my mind the best way of carrying out the colonel's views when I received a letter from a lieutenant in the navy, then on board a man of war in the Mediterranean, telling me that he had learned that the Russian sailors in the Black Sea were all supplied with revolvers, which would give them a great advantage in boarding should there be any

[1] A. V. Newton, *Colonel Colt's Small Arm Manufactory*, 1856. (Reprinted from Newton's London Journal of Arts).

naval conflict in the war then pending. This was in 1854. I at once sent this letter to the *Times*, asking if it was fair to the British sailors that they should still be limited to the old horse pistol and cutlass? The letter was published. Sir Thomas Hastings the head of the Ordnance Board at once sent for Colt and an order for four thousand pistols for the navy was given to him. This gave an impetus to the pistol trade and Colt flourished. It is due to state that he treated my little effort on his behalf with a liberality truly American.

This is a good story and in fact a letter from 'H.J.' on the above lines was printed by *The Times* on 24 February, 1854. The truth is, however that action was taken by the authorities before this date. No doubt the Admiralty were already aware of the rumours about the Russian revolvers for on 28 January the Board of Ordnance asked both Colt and Deane, Adams & Deane to submit a revolver suitable for naval use.[1] Encouraged by this, Colt's representative, R. A. Emmerson, approached another potential large buyer, the East India Company, on 21 February. Emmerson's letter was referred to their Inspector General of Stores, Col. I. J. Bonner, who dismissed it with these words: 'Revolvers are not wanted for the Public Service – I dare say the Committee have no particular wish for an Interview with Mr Emmerson for their exhibition.'[2]

However, the Board of Ordnance thought otherwise. On 6 March, Vice-Admiral Sir Charles Napier, in command of the Fleet about to sail for the Baltic Sea, sent an historic letter to the Secretary for the Admiralty:

Sir,
    I request their Lordships will be pleased to order a supply of Colts Revolvers to be made to the Fleet under my command in the following proportions viz.

Ships of the Line	100
Frigates	50
Smaller Vessels	30

I am Sir, etc.

[1] W.O. 47/2748.
[2] India Office Library, London. Records of the East India Company. Home Correspondence, Military Papers, 1854, No. 481.

The next day, the Admiralty informed him that a proportion determined upon by their Lordships had been ordered from the Board.[1] What followed is described by Sir Thomas Hastings in his evidence before the Parliamentary Committee on 14 March 1854:

> I had an interview with Colonel Colt and explained to him what was wanted and I told him that unless the arms could be furnished rapidly it would be of no avail. He then said, 'I have many demands on hand, but I will take this order and I will furnish the 4000 pistols that you want within a month and you shall have on Wednesday (the 14th.) the first 50 delivered at the Ordnance Office.' This morning a box arrived complete, with a very perfect assortment of arms and appurtenances and accompanied by a letter in which he states that in the course of a fortnight we shall have 1500 more so that when the next division of the fleet proceeds to the Baltic, these pistols which are so much wanted may be sent in the ships that are going to join Sir Charles Napier.

This interview was confirmed by the Board on 8 March when they gave Colt a proper contract for 4,000 revolvers at £2.10s.0d. each.[2] Some plated models which he had on hand were included at a price of £2.15s.0d.[3] The contract also covered the supply of percussion caps, balls and powder flasks. Colt appears to have sub-contracted for these with Eley Bros. of London and with the Birmingham firm of H. Van Wart Son & Co. of 23 Sumner Row, who were general merchants supplying the American trade.[4]

The Board, therefore, had taken the unusual step of ordering a firearm before it had been given an approved trial. This was

[1] P.R.O. Napier Papers. Entry Book of Letters to and from the Admiralty. P.R.O. 30/16/6.

[2] W.O. 47/2748.

[3] W.O. 47/2343. 6 April 1854.

[4] Henry Van Wart, the head of the firm, was an American and a brother-in-law of Washington Irving. It was probably the same man who gave evidence before the Parliamentary Committee on 23.3.1854. See also Edwards, p. 298.

arranged later in the month. On 28 March the following report was made to Joseph Wood, Secretary of the Board:

Royal Arsenal,
Woolwich
28 March 1854

Sir,

I have the honour to report, for the information of the Honourable Board, that three of Colonel Colt's revolving pistols sent to me for trial, have undergone the following tests; viz.

No. 1. Large size pistol, 400 rounds

No. 2 Second size pistol, 400 rounds

No. 3. Also second size, 150 rounds

The whole of these pistols have made good shooting at ranges of 50, 75 and 100 yards; but the accuracy of aim is of course dependent on the skill of the person who uses them.

Nos. 1 and 2 were carried in a holster by mounted N. C. officers for an hour and a quarter, trotting and walking alternately, in order to try if the bullets were secure in the chambers. They were in no way displaced, though one of the pistols exploded in the holster (slightly injuring the leg of the N. C. officer). This was occasioned by the pistol lifting up and down, there being no flounce to the holsters, to confine the pistol securely in its position.

It is necessary, in order to secure the chambers revolving freely, that the caps should fit the nipples very accurately.

I have not had the advantage of the assistance of the Sub-committee on Small Arms upon these trials, as the officers are dispersed on foreign service; but Lieutenants Warlow and L'Estrange R.A., have constantly attended.

It is our opinion upon the whole, that Colonel Colt's revolving pistols are good, effective, substantial, and serviceable arms, and with moderate care and attention would answer all the exigencies of service.

I have, etc.
J. A. Chalmer, Colonel, R.A.[1]

---

[1] I have been unable to trace this report in the records, but it is printed as Appendix No. 5 to the Parliamentary Report.

Interestingly, the accident described above was repeated the first time the revolver was used in action. When an Allied Fleet landing party destroyed the Russian fortress of Bomarsund in 1854, the first casualty was caused by Lieut. Cowell, R.E., accidentally shooting himself in the leg with his revolver.[1]

Despite Chalmer's favourable report, directed to the use of the revolver by cavalry, the Board of Ordnance were not ready to take this step. Nevertheless, it gave another order for 1,000 revolvers for the use of the Baltic Fleet on 9 June.[2] The Black Sea Fleet was the next to receive a supply and 800 were ordered for this purpose on 12 June. This figure was increased on the 16th to 2,500.[3] Colt's revolver was undoubtedly popular with the Navy and during the summer of 1854, the Admiralty was besieged with requests for them from ships' captains. It is difficult, however, to judge what was the basis of issue to each ship. The *Viper*, for instance, received 20 on 5 August and a further ten on 23 September; other ships are mentioned in the records merely as having received their 'proportion'.[4]

It has been stated that the revolvers were issued to the Royal Marines by virtue of the fact that on the grips of some existing specimens the letters RM and a Crown are stamped.[5] This mark, however, is found on many firearms, normally in conjunction with the word ENFIELD, and signifies that the gun has been through the Royal Manufactory at Enfield. At this time the Marines were issued the ordinary infantry rifle.

Colt's old rivals, Deane, Adams & Deane, now entered into active opposition. For some time they had been content with contracts for gun parts, but on 7 August they submitted a revolver to the Board, which they said they could supply at the same price as Colt's. They promised to supply 700–800 a month, stating that they were selling 200 a week to the private trade but would give all Government orders priority. Their revolver was accordingly sent to Woolwich for trial.[6]

By this time both the Adams and the Colt had their supporters

[1] P.R.O. War Office in Letter Books, Crimea. W.O. 1/386. Report of Brig. Gen. Jones, 13.8.1854.

[2] W.O. 47/2345.

[3] Ibid., 47/2749.

[4] See various entries in W.O. 47/2347 and Adm. 2/1567.

[5] Edwards, p. 306.

[6] P.R.O. In letters of the Board of Ordnance. W.O. 44/623/197.

and their detractors. Although there was little to choose between the published performances of the two revolvers, the difference in their mechanical construction made a choice between the two very much a matter of opinion. William Greener summed up the situation when he wrote in the year 1858:

> The great disadvantage said to be existing in this revolver (the Colt) is the necessity of cocking and half-cocking at every discharge; which double action is difficult in certain positions where revolvers are of the greatest use, as in a melee surrounded by many enemies where the cocking and firing by pulling motion, as in Tranter's and Dean's, is more expeditious; in fact, certificates are published by officers who, at the battle of Inkermann, would have been cut down had the slightest delay been necessary for cocking the pistol. On the other hand, it is said, that no certain aim can be taken where the pulling up and sudden liberation of the mainspring discharges the pistol; the act of discharging it destroying the aim. These two points have their advocates and their objectors, as has always been the case with new plans.[1]

At any rate, Col. Chalmer did not exactly enthuse over the Adams revolver. He fired fifty rounds through it at a target six feet square from a range of fifty yards, and although all the shots hit the target his report on 11 August finished with the words 'the machinery worked well but there was an escape of gas between the Barrel & Chamber which unless remedied would after much use be a disadvantage'.[2]

The makers were then informed that if they could remedy the fault the Board would consider trying their pistols against those of Colt. As all revolvers were subject to such a fault, no comment could have been more infuriating. They quickly arranged for two more trials; first, 100 rounds were fired and finally, in one day, 1,000 rounds were fired successfully through one revolver under the supervision of Col. Chalmer. The Duke of Newcastle, the General Commanding-in-Chief, was persuaded to write to Viscount Hardinge, the Master General, to

[1] Wm. Greener, *Gunnery in 1858* (London, 1858).
[2] W.O. 44/623/197.

ask why the American revolver was preferred to the English one. This had no immediate effect, for Col. Chalmer took the liberty of deciding that as the trials were private ones, he need make no report. It was not until 2 February of the next year that, after some pressure, his report was made.[1]

While Chalmer was thus delaying the activities of his rivals, Colt's factory at Thames Bank was a hive of industry, the stockrooms stacked to the ceilings with boxes of his naval revolvers. To get them to the fleet in time, the Board took the unprecedented step of allowing them to be dispatched direct from the factory to Portsmouth, sending down three viewers with arms chests and packers to work on the premises.[2] On 14 August a further order for 2,000 revolvers was given.[3] The Board were now well supplied with this new firearm, if short of the old ones. Rather tentatively, therefore, it issued revolvers instead of single shot pistols to a newly formed military police force known as the Mounted Staff Corps.[4] It may have been encouraged to do this by the previous issue of forty revolvers on 8 June to the Metropolitan Police Force guarding Deptford Dock Yard.[5] It was a good sign for Colt, who was anxious to get orders for supplying the Army, now that the Navy requirements had been met with satisfaction.

Anonymous letters, perhaps inspired by Colt, began to appear in the newspapers, asking why the successful revolvers were not supplied to the cavalry, etc. 'F.M.'. writing to the Editor of *The Times* on 20 November, quoted the example of the wounded Cornet Handley shooting down three Cossacks with his revolver and asked 'Surely, Sir, it might have led to important results had all the British cavalry, private as well as officers, been similarly armed in the late affair at Balaclava.' On 11 December 1854, 'An Artillery Officer' writing to the same newspaper suggested that the officers and men of his regiment should be provided with Colt's revolvers. 'It seems to me very niggardly on the part of the Ordnance not to furnish a weapon which has been proved to be so superior. Moreover, I fear it will be found that the Czar will not neglect this advantage.' On

---

[1] W.O. 44/623/197 and P.R.O. Out Letters of Secretary of State (Crimea). W.O. 6/76, p. 130.

[2] W.O. 47/2347, 21.8.1854.

[3] Ibid., 47/2750.　　　[4] Ibid., 8.9.1854.　　　[5] Ibid., 47/2749.

the same date a letter from 'A late resident in Russia' quoted an extract from a letter from St Petersburg, 'The Americans (the same who have been so much to do with the Moscow Railway) are building a great many gunboats (screws) and Colonel Colt has been or is here still with his machinery to make revolvers.'

This was followed on 19 December by another letter from 'B'.

> I have now to inform you that Colonel Colt has just returned from St Petersburg and I understand that he is about to furnish the Russian Government with a large number of his revolvers which are to be manufactured at Liége.
>
> If there be no means of preventing American and Belgian subjects from supplying our enemy with arms at least let our gallant fellows be equally as well armed as the Russians ...
>
> Away then with the antiquated pistol, carbine and lance now in use. The sword and a pair of revolving pistols are the only weapons with which our heroic cavalry and artillery should be armed.

The tone of the correspondence had taken a turn which perhaps Colt had not anticipated, and being afraid that public opinion was reaching a dangerous level, he wrote to *The Times* on 27 December from 1 Spring Gardens, complaining that since the publication of these letters he had been severely condemned. He went on:

> It is not true that I have furnished arms or machinery to the Russian Government or that I have contracted to furnish either arms or machinery to that Government. The only truth in the letter, as regards me, is that I have been in Russia as I have been in the other great States of Europe, during the last two months. Ever since my armoury has been established in London, both it and my own skill have always been at the service of the Government and it rests with them to employ either or both to their fullest extent. My offers to the Government to manufacture any description of arms at prices much less than are paid to others, have been sufficiently public already and should afford a complete answer to all

complaints against me of the sort referred to. It is not my fault if all my facilities are not now devoted to the British Government.

It was all publicity, however, and the year ended successfully with an order for 5,000 revolvers at £2.10s.0d. each and an unexpected issue of 20 revolvers to the Director of Convict Prisons at Dublin.[1] Better still, from Colt's point of view, the large order was destined for the Army in the Crimea. The Duke of Newcastle wrote to Lord Raglan on 4 January 1855:

> I have the honour to acquaint your Lordship that in pursuance of directions from myself, Colonel Colt has signified to the Board of Ordnance his readiness to supply 3000 Revolving Pistols immediately for the use of the Army under your Command and 2000 in addition to that number by the 12th. February.
>
> It is my intention in ordering these Arms to be placed at your disposal that they should be used either by the Cavalry or by a Storming Party or by Colour Serjeants or in any way which your Lordship may consider most desirable.[2]

By 13 January 1,000 of these had been inspected and were ready for packing.[3] Colt was well satisfied with the position and returned to the U.S.A. There he was met by the British arms representative Capt. Marlow, R.A., who had been intructed to buy three drop hammers from him.[4]

With Colt's revolvers forming a major item in their arms stores, the Board of Ordnance had to face various problems connected with its accessories. The main difficulty was to design a proper holster and belt. On 30 January 1855 the Board ordered a special 'Frog' for use of the Navy, to carry the revolver attached to the sword belt. Provision was at first made for a powder flask but this was later discarded.[5] The Army does not seem to have been so well served, and there were some complaints that the revolvers given to the sergeants were

---

[1] W.O. 47/2751, 23.12.1854, 26.12.1854, 28.12.1854.

[2] W.O. 1/385.

[3] W.O. 6/78, p. 145.

[4] W.O. 47/2752. Minute of 16.2.1855.

[5] W.O. 47/2352 (30.1.1855); W.O. 47/2353 (22.2.1855).

of little use, as they had no means of carrying them properly.[1] On 5 May, cartridges designed by Col. Maher, R.A., were sent to Capt. Maitland at Portsmouth for trial with the Colt, and at the same time some of Colt's own cartridges, presumably the waterproof tin foil type he patented at the end of the year, were also submitted.[2] Bills of a later date indicate that the latter were adopted.[3]

An interesting situation had now arisen. Gunmakers all over the country, working in countless small workshops, were striving to fulfil their contracts and failing. Their combined efforts could hardly keep pace with Colt's single factory. On 15 February 1855 the Ordnance was forced to place an order for 25,000 rifles in the U.S., and gave an advance payment of £21,875 to the agents Fox & Henderson.[4] But the arms shortage continued. The Board began to issue revolvers wherever it could. In May the Land Transport Corps, composed mainly of native drivers, were issued Colt's revolvers.[5] On 7 June the stock of ordinary pistols was exhausted and the Board received the reluctant permission of Viscount Hardinge to issue revolvers instead, first to the 12th Lancers and then to the 17th.[6] On 2 August Colt received his largest order, consisting of 9,000 revolvers.[7]

In spite of this, prospects for the future were not so good. For one thing Colt's competitors were now well in the field. In March, Lieut. Beaumont had supplied 100 of his revolvers to the Tower and on 1 June 1855 the Ordnance Office sent one of these to Capt. Maitland for his approval with the instruction that if approved, they would be adopted in future for the Navy.[8] Deane, Adams & Deane had started in July a monthly supply of revolvers[9] and although Colt knew they were unlikely to

---

[1] R. H. Murray, *The History of the VIII King's Royal Irish Hussars* (Cambridge, 1928), p. 430.

[2] Adm 2/1570; W.O. 47/2753 (11.5.1855).

[3] W.O. 47/2754, 19 and 26 September 1855.

[4] W.O. 47/2752.

[5] P.R.O. Out Letters, Commander in Chief, W.O. 3/325. Additional issue of 116 Colts revolvers made on 9.5.1855. On 29 September 1855, one of these was reported to have 'Burst on discharge' (W.O. 3/118).

[6] W.O. 3/325. (7.6.1855, 4.8.1855).

[7] W.O. 47/2754.

[8] Adm. 2/1570.

[9] W.O. 47/2754. They made a claim for payment 9.7.1855.

match his production, the fact remained that he had lost his monopoly. Unless a wider issue of revolvers was made, he was unlikely to get orders of the size he needed to keep his factory going. The crucial stage had been reached.

The stumbling block to both Colt's hopes and the Board's plans was Viscount Hardinge. He was fascinated with the new breechloading carbines that were being offered to the Ordnance by both English and American inventors. On 8 August, when the Superintendent of the Enfield factory told him that the stock of Victoria carbines was low and suggested that revolvers should be issued to the cavalry, he replied, 'Colt's Revolving Pistols may as already authorized in the case of the Depots 12 & 17 Lancers be supplied to Regiments of Lancers but they are not adapted for the Cavalry generally.' Instead he strongly urged the Ordnance to adopt Leetch's breechloading rifle.[1]

For the next few months Colt was occupied with supplying the 9,000 revolvers, and on 6 October he won the contract for the full set of accessories to go with them.[2] Most of these revolvers were sent out to the Crimea. On 20 August, 1,500 were sent to the Ordnance depot at Balaclava for the use of General Simpson, who was directed to distribute them equally to the Company Officers and Sergeant Majors of the Infantry regiments.[3] Some of them, however, found their way to strange homes. A schedule of plant belonging to the Crimean Railway Expedition at Balaclava in 1855 included: 3 cases Revolver Pistols; 3 cases Belts; 3 cases ammunition; 1 case 30 packages 50 rounds each.[4]

England was also supplying arms to the various foreign legions under its control and on 28 December 1855 there was an issue of 500 revolvers to the two cavalry regiments belonging to the 'Cossacks of the Sultan' under Count Zamoyski.[5] The Colt was a cosmopolitan weapon!

At the beginning of 1856, with the end of the Crimean War in

[1] W.O. 3/325.

[2] W.O. 47/2755.

[3] W.O. 3/118. Adj. Gen. to General Simpson.

[4] W.O. 1/378.

[5] W.O. 6/74, p. 139/145. This was actually a Polish regiment, but for various political reasons it was classified as a Turkish Corps d'Armee.

sight, there was a poor outlook for the revolver makers. For the first few months, Colt continued to send in spare parts – screws, springs, nipples, etc., but all interest was now centred on the breechloaders. As well as Leetch, the London gun-maker Prince and the two Americans Greene and Sharps were trying to get their carbines adopted. No new contracts were issued for revolvers but an order for 6,000 Sharps carbines was given on 7 July 1856.[1]

Charles Dennet made another attempt to interest the East India Company in the revolvers and wrote a long letter to the Chairman, Lt.-Col. Sykes, on 22 May 1856. He said that about 35,000 had been supplied to the Army and Navy,[2] but that he had about 5,000 in hand which he could supply 'on the most favourable terms'. He added the following note:

> The introduction of the cartridges does away with the necessity of a Flask – The new Catouche Pouche sent carries both the Cartridge & Caps & is considered a great desider-atum – The 9000 Turkish Contingent were supplied with them.[3] The belt might be done away with, as the Holster and Catouche Pouche – would go on the sword belt. The bullet mould too – need not be furnished. As we have a Mould casting 12 at once – Cutting down the parts and appurtenances as sold at retail – at £6.10.0 – makes the weapon cost in gross quantities very considerably under £2.15.0.
>
> I have voucher of recent trial in India where, at eighty yards, the Pistol beat the Carbine.

Bonner's reply was short but final:

> Under the new arrangement with Her Majesty's Govern-ment all Fire Arms of every description are obtained from

---

[1] W.O. 47/2758.

[2] It is difficult to judge the accuracy of this statement for it contradicts others made by Colt. The figures given in this article, taken from official records which appear to be complete, give a total of 9,500 to the Navy and 14,000 to the Army.

[3] On 21.2.1856 Dennet and an Army contractor, Geo. Pays of 260 Oxford St., London, took out a Provisional Protection under British Pat. No. 451 for some 'Improvements in Cartouche and Percussion Cap Pouches'.

the Head Stores of the War Dept., the East India Company making no direct provision whatever.[1]

With such a large stock in hand, work in Colt's factory began to slow down. On 14 November, the last payment from the Ordnance was made – for bullets surplus to contract.[2] Colt, writing home from St Petersburg, described the situation:

> We have accumulated a large surplus of arms in the London Armoury. I have ordered that no more new work shall be put into the works there. Consequently several of our Yankee boys have gone home & some of them may apply to you. If there is a place you must give them employment for I do not want to loose sight of them in case they are again wanted in Europe . . . [3]

However, Europe was set on peace – at least for a short while, and Colt began the dismal process of the disposal of his factory. It was ironical that at this moment he was made aware that he had been occupying the factory without a proper lease. The Office of Works and their landlords, the Commissioners of Woods and Forests, had only just reached agreement. On 16 December 1856 a copy of the underlease was sent to Colt, and as soon as it was engrossed, in March 1857, he applied for permission to transfer it to a Mr Judson, an India rubber manufacturer. This, however, was refused for a new tenant had his eye on the premises; none other than the War Office. The two Government departments conspired together and, after a short negotiation during which he was in no position to dictate terms, Colt thankfully gave up his lease. In 1858, the Thames Workshop became a Government Small Arms Repair Establishment and a training school for Armourer Sergeants.[4]

Almost as a side line, Colt had taken out a patent in 1856 for a bullet lubricator and a backsight, and it was with these accessories, in an effort to revive their failing fortunes in London, that Dennet once more wrote to Col. Sykes of the East India Company on 2 June 1855.

[1] E. I. Coy. Home Corres. Military Papers 1856. No. 554.
[2] W.O. 47/2759.
[3] Edwards, p. 321.
[4] Works 6/160/9 and Letter Books, Works 1/54.

Col. Sam Colt's Offices
14 Pall Mall
June 2nd. 1857

Sir,

I beg to enclose a specimen of a new invention by Col. Colt and also reports of experiments & trial thereof by the U.S. Government.

The Lubricator can be made applicable to any species of Fire-Arm – of any bore. And I am ready to contract for the supply of the article, like the sample enclosed for the Enfield Rifle Musket, at 2/6d. each.[1]

May I beg the favour of an inspection & trial by the Bd of Directors of the Hon E. I. Co. & to respectfully solicit an order for the same.

I beg also to forward a specimen of a new cartridge for our Revolver – which have been found to be a great improvement on any other made   I remain, Sir,
                                      Yr Very Obt Servt.
                                      Chs. F. Dennet
                                      Agt. for Col. Colt.

Colonel Bonner, however, rebuffed him again, referring him to the War Department with regard to the Lubricator, and pointing out that as the Colt revolver had not been introduced into their service, the Company had no need for the new cartridge.[2]

When the Government took over his factory, Colt found other premises (besides his Pall Mall office), at 37 Chandos Street, at the back of the Strand. The London Directories describe this as a 'Patent Repeating Rifle Manufactory'. It is possible that Colt tried to compete against the breechloaders with his revolving rifles but there is no evidence to support this. What is more likely is that he removed the remainder of his stock and some of his machinery to this address. In 1861, the Directories show that he had left these premises too. His venture in London had come to an end.

[1] The Lubricator and the patent backsight were fitted experimentally to the Enfield Rifle (Pat. 1853), but were not adopted as a general issue.
[2] E. I. Coy Home Corres. Mil. Papers 1857. No. 589.

Bibliography

This bibliography is simply a list of relevant books and articles published since Professor Habakkuk's work appeared. Students who wish to go into the matter in greater detail should read carefully the footnotes to that book, to the articles reprinted here and consult the readings suggested in the collection of essays edited by Aldcroft, the first of the works listed below.

D. H. ALDCROFT, (ed), *The Development of British Industries and Foreign Competition, 1875–1914* (University of Toronto Press, Toronto, 1968).

E. A. BATTISON, 'Eli Whitney and the Milling Machine', *Smithsonian Journal of History*, I (1966).

M. BLAUG, 'Survey of the Theory of Process Innovations', *Economica*, XXX (1963).

P. DAVID, 'The Mechanization of Reaping in the Ante-Bellum West', in H. Rosovsky, *Industrialization in Two Systems* (New York, 1966).

R. W. FOGEL, 'The Specification Problem in Economic History', *Journal of Economic History*, XXVII (1967).

H. J. HABAKKUK, 'Second Thoughts on American and British Technology in the Nineteenth Century', *Business Archives and History*, III (1963).

D. S. LANDES, 'Factor Costs and Demand: Determinants of Economic Growth', *Business History*, VII (1965).

D. S. LANDES, *The Unbound Prometheus* (New York, 1969).

S. LEBERGOTT, *Manpower in Economic Growth* (New York, 1964).

D. N. MCCLOSKEY, 'Productivity Change in British Pig Iron, 1870–1939', *Quarterly Journal of Economics*, LXXXII (1968).

P. E. MCGOULDRICK, *New England Textiles in the Nineteenth Century* (London, 1968).

E. MANSFIELD, *Industrial Research and Technological Innovation* (New York, 1968).

J. J. MURPHY, 'Entrepreneurship and the Establishment of the American Clock Industry', *Journal of Economic History*, XXVI (1966).

W. D. NORDHAUS, *Invention, Growth and Welfare* (Massachusetts Institute of Technology, 1969).

N. ROSENBERG, 'Technological Change in the Machine Tool Industry, 1840–1910', *Journal of Economic History*, XXIII (1963).

N. ROSENBERG, 'Anglo-American Wage Differences in the 1820s', *Journal of Economic History*, XXVII (1967).

N. ROSENBERG (ed), *The American System of Manufactures* (Chicago, Ill., 1969).

N. ROSENBERG, 'The Direction of Technological Change', *Economic Development and Cultural Change* (1969).

W. E. G. SALTER, *Productivity and Technological Change* (Cambridge, 1960).

S. B. SAUL, 'The Machine Tool Industry in Britain to 1914', *Business History*, X (1968).

S. B. SAUL, *The Myth of the Great Depression* (London, 1969).

P. TEMIN, *Iron and Steel in Nineteenth Century America* (Massachusetts Institute of Technology, 1964).

P. TEMIN, 'The Relative Decline of the British Steel Industry, 1880–1914', in H. Rosovsky, *Industrialization in Two Systems* (New York, 1966).

P. TEMIN, 'Labor Scarcity and the Problem of American Industrial Efficiency in the 1850s', *Journal of Economic History*, XXVI (1966).

P. TEMIN, 'Steam and Waterpower in the Early Nineteenth Century', *Journal of Economic History*, XXVI (1966).

D. WHITEHEAD, 'American and British Technology in the Nineteenth Century', *Business Archives and History*, III (1963).